# FOREST VINES TO SNOW TUSSOCKS

# FOREST VINES *to* SNOW TUSSOCKS

THE STORY OF NEW ZEALAND PLANTS

**John Dawson**

**Victoria University Press**

VICTORIA UNIVERSITY PRESS
Victoria University of Wellington
Private Bag Wellington


First published 1988
Reprinted with corrections 1993
ISBN 0 86473 047 0



National Library of New Zealand
Cataloguing-in-Publication data

DAWSON, John
Forest vines to snow tussocks: the story of New Zealand plants/John Dawson. — Wellington (N.Z.): Victoria University Press, 1988. — 1 v.
ISBN 0-86473-047-0
581.9931
1. Botany — New Zealand. I. Title.

Typeset by Computype Ltd, Wellington
Printed in Hong Kong through Bookprint
Consultants Ltd, Wellington

# Contents

# Foreword

Dr John Dawson's *Forest Vines to Snow Tussocks* should be welcomed by all those interested in the plants of New Zealand, and indeed in the plants of the south-west Pacific and the far South. Earlier generations could turn to Leonard Cockayne's *New Zealand Plants and their Story* or Robert Laing and Edith Blackwell's *Plants of New Zealand* for introductory accounts of the vegetation of New Zealand, the origin of the flora, or discussions about the evolution of our plants. But these works have been out of print for several years; and in any case they were written well before the Second World War. Since then there has been a remarkable expansion of our knowledge of the plants of New Zealand, an expansion to which Dr Dawson himself has contributed; and in the following pages he has brought together much of this new information, hitherto scattered in the learned journals, as well as adding insights of his own. This new book, then, fills an important gap in the literature, and will be studied with advantage by students of many kinds; and it will help to maintain interest in the preservation of the native plants and vegetation about which Dr Dawson writes so well.

Eric Godley
lately Director, Botany Division,
DSIR, Christchurch

Dedicated to
HUGH DOUGLAS GORDON
Professor of Botany, Victoria University
1947 – 1977

# Preface

It is now 60 years since the third edition of Cockayne's *New Zealand Plants and their Story* was published. A fourth edition, edited by E. J. Godley, appeared in 1967 with the text virtually unchanged, but with updated botanical names and references in footnotes to more recent information. In the preface to this edition Godley says: 'Dr Leonard Cockayne died in Wellington on 8 July 1934 at the age of 79, seven years after the third edition of this book appeared. The publication of a fourth edition would probably have pleased him, in showing that his story was not forgotten and in making it again widely available. But I think too that he would have criticised us — for he was a forthright man — for not having produced something to take its place.'

The aim of the present book is not to replace Cockayne's account, which, as Godley states, is 'a classic of New Zealand botanical literature', but to build on it by reviewing the increased knowledge of New Zealand plants and their communities, living and fossil, gained over the past 60 years. As with *New Zealand Plants and their Story*, identification of species is not the primary purpose of this book. A number of guides to the identification of various groups — ferns, trees and shrubs, alpines - already meet that need. The emphasis is on a representative selection of species, their life styles, their interesting and peculiar features, their histories and relationships, presented within the framework of their communities. In addition, general characteristics and peculiarities of the flora and theories relating to them are reviewed.

As is the case with most general accounts of the floras of countries, this book is chiefly concerned with the larger and more conspicuous vascular plants — ferns and their relatives, conifers and flowering plants. The generally smaller mosses, liverworts, lichens, fungi and seaweeds are equally numerous and diverse in New Zealand, but that is a story that others are better qualified to tell.

I hope that this book will be useful to the interested layperson, to students and to botanists both in New Zealand and overseas. I have tried to avoid technical terms as much as possible, but the use of botanical names for species, as well as common names where these are well-established, seemed unavoidable. The overseas botanist is unlikely to find 'supplejack' or 'bush lawyer' very meaningful, but *Ripogonum scandens* and *Rubus cissoides* will convey a great deal about our species from features shared with their relatives elsewhere.

The book derives in large part from courses given at both graduate and undergraduate levels and also over a number of years to Workers' Educational Association classes.

## Acknowledgements

I am very grateful to a number of botanists who have made useful suggestions on the whole manuscript or parts of it. E. J. Godley, B. V. Sneddon, W. R. Philipson and B. P. J. Molloy reviewed the entire manuscript; F. B. Sampson, D. R. McQueen and G. M. Rogers read some early chapters; P. S. Green, P. Williams, W. R. Sykes and J. Campbell advised on Chapter 10 and D. Pocknall, D. Mildenhall and A. M. Holden on Chapter 12.

For technical assistance I am especially grateful to Maureen Cooper for word processing the several revisions, to John Casey, his predecessor M. D. King and the Photographic Facility for taking some of the photographs and for printing most of the others, to Philippa Spackman for preparing several maps and diagrams, and to the editorial staff of Victoria University Press.

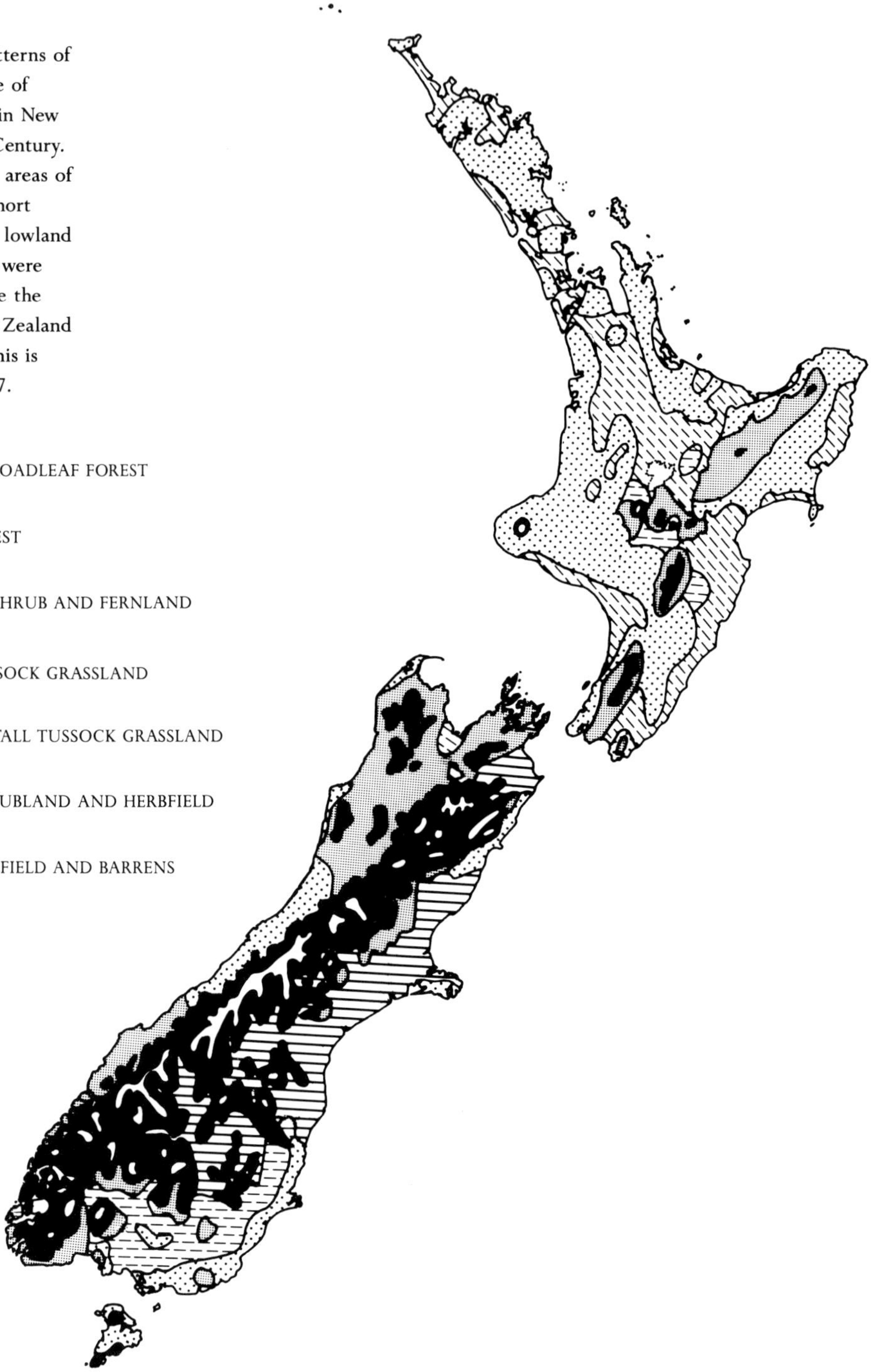


*Figure 1* The broad patterns of vegetation at the time of European settlement in New Zealand in the 19th Century. It is believed that the areas of fern and shrubland, short tussock grassland and lowland tall tussock grassland were largely forested before the Maori settled in New Zealand 1000 years earlier. This is discussed in Chapter 7.

CHAPTER 1

# General Characteristics of the New Zealand Flora

The current estimate for the number of vascular plant species native to New Zealand is about 2200.[1] This is a modest total by comparison with tropical floras, but it is of moderate size for a temperate country. About 82 per cent of the species are endemic; a high proportion reflecting New Zealand's isolation.

New Zealand is a generally moist country with mild temperatures in the lowlands. It is believed that before human habitation much of the landscape was covered by dense forests — conifer broadleaf[2] forest predominating in the warmer North Island and parts of the South Island and beech (*Nothofagus*) forest predominating in the South Island and, mostly at higher altitudes, in the North Island. Human activities have greatly reduced this forest cover. Above treeline, ranging north to south from about 1500–1000 m in altitude, there are various types of short alpine vegetation, which sometimes include a belt of shrubs as a transition from the forest (Fig. 1).

Of the world's larger islands, New Zealand's narrowly separated pair are the most remote from any continent. The British Isles and Japan are as close to their adjacent continents, at their nearest points, as the North and South Islands are to each other; the larger islands of Indonesia and the Caribbean form close set series between continents and even Madagascar is, at 400 kilometres from Africa, only one quarter the distance from that continent that New Zealand is from Australia, its nearest continental neighbour.

Most of New Zealand's rocks are sedimentary, laid down originally beneath the sea, and as a basis for our understanding of the history and characteristics of the flora we need to know whether New Zealand emerged in its present isolation or whether it was first connected with a larger landmass.

Before considering what geologists have to say on this point we will see whether the plants themselves can tell us anything. Do floras which have evolved on isolated islands which have never been connected to a continent have any characteristics to distinguish them from continental floras? Studies of both the flora and fauna of Hawai'i,[3] as well as of other isolated islands,[4] suggest that the answer to this question is 'yes'.

## The Isolated Island Syndrome[5,6]

The area of the Hawaiian islands is much less than that of New Zealand but their isolation is much greater, since the nearest continent, North America, is 4000 km distant and both Australasia and Asia are about 7000 km distant. The Hawaiian islands are entirely volcanic and have arisen sequentially on a north-west to south-east line over the last 4–5 million years. Hawai'i, the southernmost island, is thus the youngest and is also the largest and highest (4000 m). There seems little possibility, geologically, that these islands have ever been connected to a continent, so botanists and zoologists alike envisage their being stocked by chance immigrants with subsequent evolution of new species. Thus it has been estimated that 272 immigrant species could have evolved into the present approximately 1200 native higher plant species of Hawai'i. The distinguishing characteristics of floras that have been derived in this way can be outlined as follows:

### Getting There

Not unexpectedly, plant and animal groups with good dispersal ability are strongly represented, while those with poor dispersal ability are absent. Among plants, those with seeds that have special modifications for dispersal (for example: flotation devices for sea transport; plumes of hairs for wind dispersal; hooks for attachment to birds' feathers; or hard seeds in berry fruits eaten by birds and eventually excreted intact), have a good chance of eventually reaching an isolated island. Plant groups which are notable for not reaching such islands, because their seeds are large and/or unspecialised, include conifers; the wind pollinated trees (including oaks and beeches), so important in some temperate forests of the northern hemisphere; and most of the primitive woody flowering plant families in the order Ranales, of which *Magnolia* is the best known example. Prominent among animals on isolated islands are those with wings — birds and insects — while mammals, amphibians and land reptiles are absent.

Strangely, many species of small, isolated islands have lost or have only vestiges of the dispersal mechanisms possessed by their continental relatives. Thus there are many flightless island birds and insects. Among plants, species belonging to groups that are normally wind dispersed may have seeds with quite inadequate vestigial wings or plumes of hairs. In explanation of this it has been suggested that, although a good dispersal mechanism is necessary for a plant species to reach an isolated island, once the plant is established, good dispersability would become a disadvantage for most of the seeds would end up uselessly in the sea!

### Adaptive Radiation

Many genera on isolated islands exhibit a much wider range of growth forms and occupy a much wider range of habitats than the same or related genera on continents. Thus on mountainous islands like Hawai'i, some species of a genus may be small and herbaceous and occupy coastal and lowland open habitats, other species may be small trees in wet forests at various altitudes, and yet others shrubs or herbs in alpine vegetation. Carlquist,[3,4] who has closely studied island floras and faunas, suggests that the first species of a plant genus to arrive is likely to be a weedy herb, because in general weeds are good dispersers and hardy enough to tolerate the raw open conditions of a new volcanic island. In such a situation many habitats will remain unoccupied for some time (particularly, Carlquist suggests, those suitable for moist forests), so any variants of the weedy coloniser with habitat requirements different from the parent population would have a very good chance of establishing in an unoccupied niche. In turn a variant population could give rise to other variants to occupy further niches including, if the island is high enough, those of alpine habitats. Such a burst of speciation into unoccupied habitats has been termed 'adaptive radiation'. It is not suggested, however, that a species increases its ability to give rise to new varieties and species on migrating from a continent to an isolated island. On the continent it would also form variants, but these would rarely find unoccupied habitats suited to their needs.

Carlquist's view that weedy herbs are frequently the first colonisers of isolated islands implies a general trend in plant evolution on some islands from herbs to trees. This is entirely possible, but it would be an unusual direction for evolution to take, if the generally accepted idea that herbs are advanced, relatively recent in origin and derived from woody ancestors, is correct. In opposition to Carlquist's view some botanists believe that tree species on isolated islands, belonging to genera or even families which are otherwise mostly herbaceous, are primitive ancestral forms that have survived there because of the mild oceanic climates and reduced competition.[7] On continents, particularly those of the northern hemisphere, woody species of these groups have generally given way to herbaceous species better suited to the strongly seasonal climates which have developed at higher latitudes on continents in recent geological time.

**Sexual Patterns**

It would seem obvious that a plant species with self-fertile hermaphrodite flowers would stand a better chance of establishing on an isolated island than an hermaphrodite species which is self-sterile or a species with separate male and female plants, a condition known as dioecism. In both the latter cases one plant would not be sufficient to form a population; at least two would be required — one male and one female in the case of a dioecious species — and they would have to establish at the same place and coincide or overlap in time. In some cases, however, this might not be unlikely, as in some species seeds tend to be transported in groups. Berries often contain several small seeds, so a bird might eat a number of berries of a particular species, transport them internally and deposit them in a group on an isolated island. This could be the case with the dioecious, berry-fruited genus *Coprosma*, so strongly represented in both the New Zealand and Hawaiian floras. Sticky or barbed seeds which attach themselves externally to birds or seeds in mud might also be transported in groups, but this would be less likely with seeds conveyed by winds or ocean currents. Nevertheless, the odds would seem to favour self-fertilising plants as colonisers of isolated islands, and we would expect a lower percentage of say dioecious species on such islands than on a continent. Paradoxically, the reverse is the case. In Hawai'i it is estimated that 27.5 per cent of the species are dioecious,[3] a much higher proportion than for the United Kingdom, for example, where it is estimated only 2 per cent of species are in this category.

How can this be? If dioecious species are less likely to colonise isolated islands and yet are found there in relatively large numbers, there must be some circumstance which especially favours both the survival and diversification of the relatively few dioecious colonisers and the evolution of dioecious from hermaphrodite island species. This implies that dioecism confers some special advantage in the isolated island situation. It has been suggested that this advantage results from the obligate outcrossing of dioecious species, which brings about an increase in variability. Variable species are more likely to be successful on isolated islands as they are in a better position to take advantage of the unoccupied habitats available, and more likely to survive drastic environmental changes such as those resulting from the succession of glacials and interglacials in recent geological time. On a continent stretching from high to low latitudes, species can survive such fluctuations by migrating with their preferred climates as these climates move towards and away from the equator. On an isolated island, this option is greatly restricted, so with a drastic environmental change, a greater proportion of relatively invariable inbreeding species will become extinct than of more variable outcrossing species, which are better able to adapt to change.

## Hybrids

A similar explanation has been suggested for the high level of natural hybridism on isolated islands.[8,9] The progeny from such hybrids displays an even greater variability than that resulting from dioecism, and should thus be an even richer source for the populating of unoccupied habitats or new habitats resulting from environmental change.

## Lack of Brightly Coloured Flowers

On islands like Hawai‘i an unusually high proportion of the native plants have small, shallow flowers which lack bright colours and are pollinated by wind or unspecialised short tongued insects. The exceptions are certain bird-pollinated species which have larger red to yellow flowers of tubular form.

The general lack of brightly coloured flowers is attributable to the absence from isolated islands of the more specialised insect pollinators, particularly long tongued bees, which are attracted by bright colours and pleasant perfumes.

## The New Zealand Pattern

Such then is the 'isolated island syndrome'. The question now arises: 'does the New Zealand flora also conform to this pattern?' The answer this time is 'yes and no'.

The strongest difference from the isolated island pattern lies in the presence in New Zealand, mainly in forests, of plants belonging to groups of poor dispersal ability.

(a) Conifers are a prominent feature of the forests with the giant kauri (*Agathis australis*) in the far north and the more widespread species of the family Podocarpaceae ('Podocarps') including rimu (*Dacrydium cupressinum*), kahikatea (*Dacrycarpus dacrydioides*) and totara (*Podocarpus totara*).

(b) The largely northern temperate group of wind-pollinated trees is represented in New Zealand by species of southern hemisphere beech (*Nothofagus*) which form extensive forests at mostly higher altitudes and latitudes.

(c) The woody families of the order Ranales are represented in New Zealand by several species, including horopito (*Pseudowintera* species), tawa (*Beilschmiedia tawa*) and pukatea (*Laurelia novae-zelandiae*).

The presence of representatives of these and other ancient and mostly southern groups not represented on islands which have always been isolated, indicates that New Zealand has not always been as isolated as it is now and must once have had continental connections.

How, when and where was New Zealand linked to other lands to receive this ancient element? When speculating on this point, proponents of the 'land-bridge' theory have made much of the fact that New Zealand is situated on an extensive system of relatively shallow submarine plateaus and ridges — the Chatham Rise and the Campbell Plateau to the south-east extending to about 55°S and the Lord Howe Rise and Norfolk Ridge to the north-west extending to about 20°S (Fig. 2). Recent evidence suggests that these plateaus and ridges are composed of continental rocks and were once land. New Caledonia, at the northern end of the Norfolk Ridge, has rocks very similar to those of New Zealand and has an ancient southern element in its flora that is related to but even larger than that of New Zealand. Thus we have to consider past connections, not just of New Zealand, but of a small and now largely submerged continental landmass of which New Zealand and

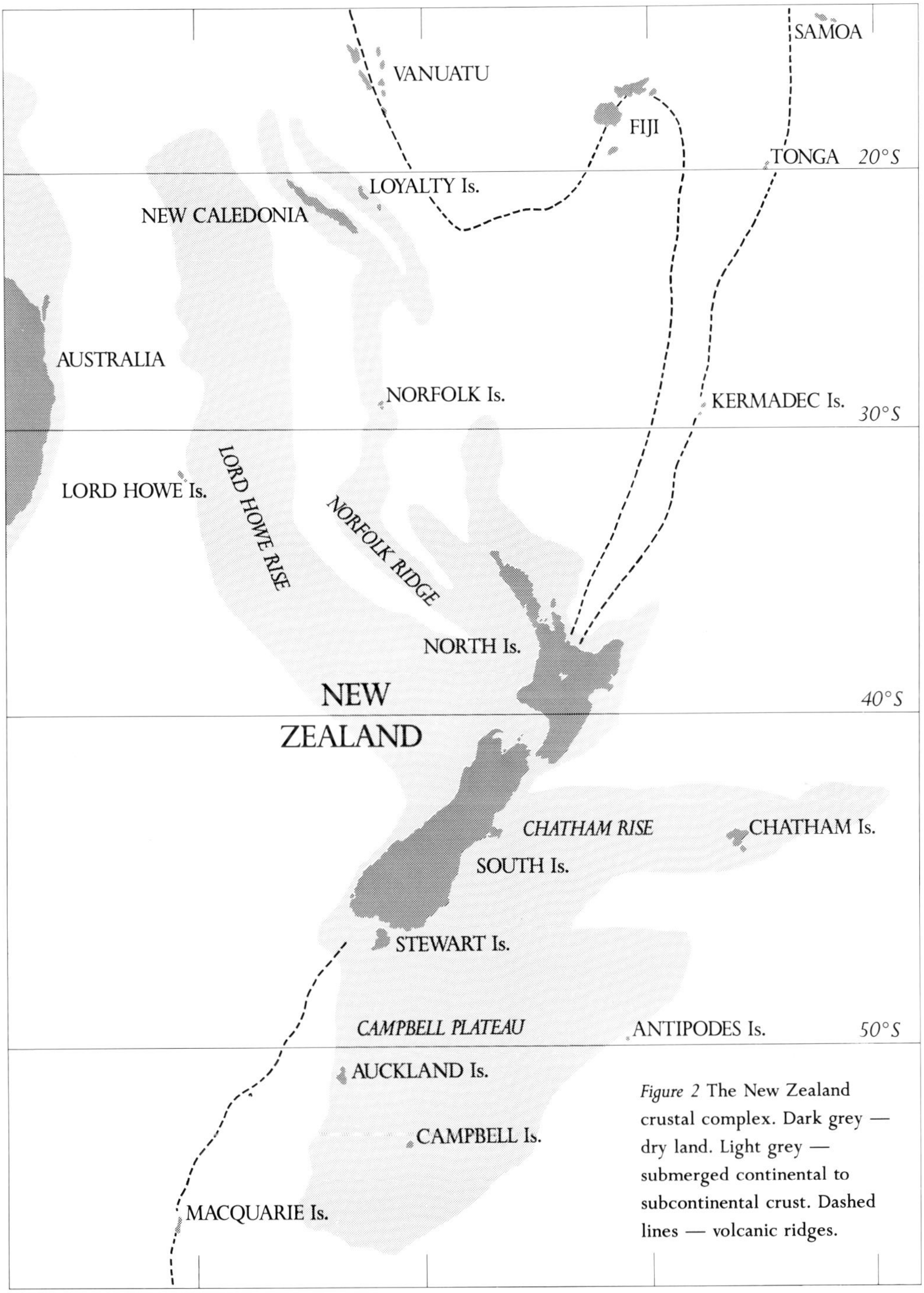


*Figure 2* The New Zealand crustal complex. Dark grey — dry land. Light grey — submerged continental to subcontinental crust. Dashed lines — volcanic ridges.

New Caledonia are the only parts still above sea level. For convenience of reference this small continent could be termed Tasmantis, a name that has earlier been applied to a hypothetical land in the Tasman Sea region.

According to the land-bridge theory, then, if Tasmantis were once largely above the sea then dry land may have extended to New Guinea and perhaps even to south-east Asia. Such a land extension might explain some of our floristic links with the tropics, but not so readily those with the former flora of Antarctica and the present flora of South America, as the 2000 km gap between the Campbell Plateau and Antarctica makes it difficult to envisage a former dry land connection between them.

Another theory provides a more likely explanation. Recent geophysical evidence of sea floor spreading has led to a fairly general acceptance of the old theory of continental drift, with some modification.[10] As far as the southern hemisphere is concerned it is believed that all its land areas, as well as India, were once united in a single large continent known as Gondwana. India and Africa/South America first separated off and later became separated from each other. Through its southern extremity South America probably retained tenuous links with the Antarctica/Australia/Tasmantis remnant of Gondwana (Fig. 3). It is suggested that Tasmantis separated and moved into isolation about 80 million years ago and that Australia and Antarctica fully separated about 50 million years ago. If the timing of these separations is correct then the ancient southern floral element could have reached New Zealand and New Caledonia before Tasmantis drifted into isolation. The presence of this element in the New Zealand flora is the most notable contrast with the floras of isolated islands, but in other respects there are several points of agreement:

(a) A number of genera in New Zealand, like many genera of isolated islands, exhibit a wide morphological and ecological range. To cite *Coprosma* again, some species are small trees with large leaves in the lowland forests, others are densely branched small-leaved shrubs mostly of open habitats from the sea coast to above tree-line in the mountains, and a few are mat-forming, near-herbs of the high mountains and montane riverbeds. Similar wide ranges are evident in a number of other genera including *Pittosporum* and *Myrsine*. Even in genera with only a few species in New Zealand there may be an

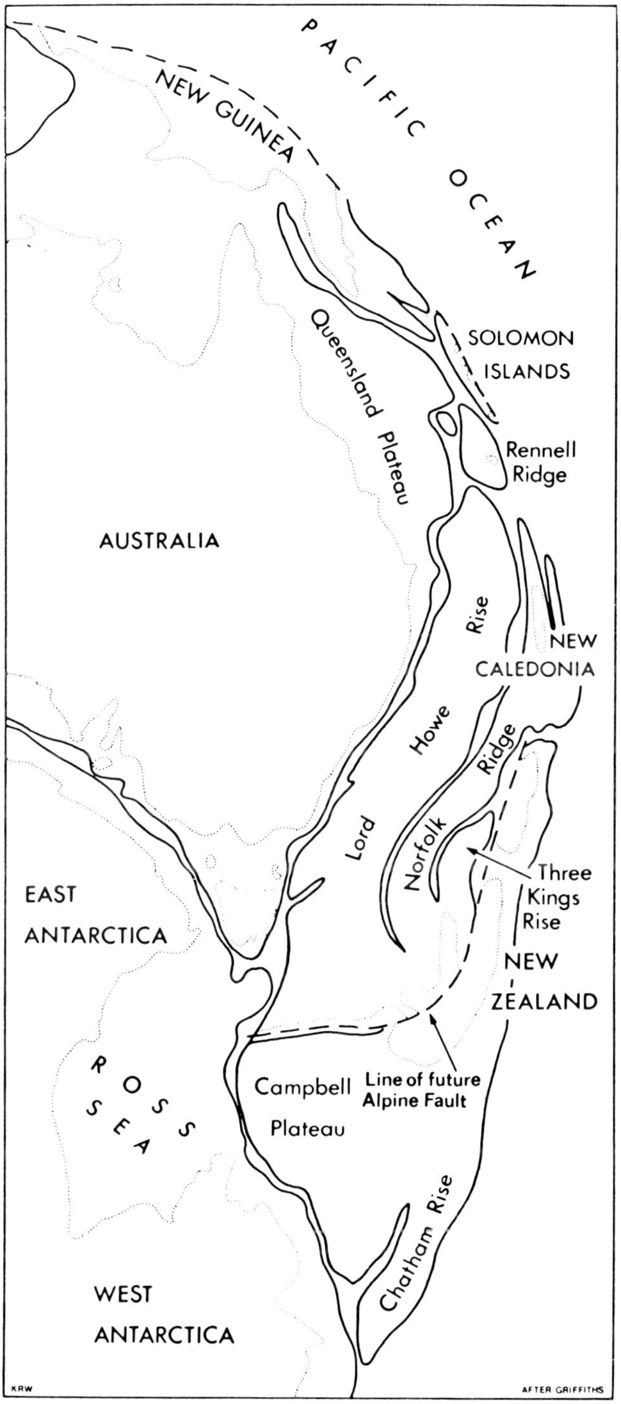


*Figure 3* Proposed reconstruction of the Australian/Antarctic/Tasmantis portion of Gondwana during the mid-Cretaceous about 90 million years ago. After Tasmantis separated lateral movements along the alpine fault gradually rearranged the crust of present New Zealand.

*Figure 4* Kaikomako (*Pennantia corymbosa*): A New Zealand example of dioecism. The female flower on the left has a well developed ovary at the centre, but the small stamens do not produce viable pollen. The male flower on the right has only a rudimentary non-functional ovary, not discernible in this photo, but has large stamens with long filaments. Kaikomako is probably wind-pollinated. Photo: B. V. Sneddon.

extreme range in growth habit. Of our three species of *Fuchsia* one (*F. excorticata*) is a small tree, one is a scrambling climber (*F. perscandens*) and the last (*F. procumbens*) is a prostrate, spreading near-herb.
(b) The level of dioecism in the New Zealand flora (Fig. 4) is, at 12 per cent,[11] much lower than that of Hawai'i, but still much higher than that of Europe. In particular some genera shared with Europe, such as *Clematis* and *Rubus*, are largely dioecious in New Zealand and predominantly hermaphrodite in Europe. In the Umbelliferae (carrot family) of New Zealand about 80 per cent of the species are dioecious or gynodioecious (female and hermaphrodite plants). Elsewhere in this large family such sexual patterns are very rare. Recently it has been shown that some tropical forests can have a level of dioecism between 20 and 30 per cent.[12] However, affinities with tropical dioecious plants would offer an explanation for only a minority of New Zealand's dioecious species.
(c) Natural hybridism has also long been recognised as a feature of

the New Zealand flora,[13] in many cases involving species of widely different form and ecology.

(d) With the exceptions of the brightly coloured, bird-pollinated flowers such as the ratas (*Metrosideros*) and kowhais (*Sophora*), New Zealand plants on the main islands, like those of Hawai'i, are not notable for size or colour of flowers. Even among alpine plants, New Zealand species of genera which may be colourful elsewhere (*Ranunculus, Myosotis, Gentiana*) are often white. Here too the suggested explanation is a lack of specialised insect pollinators.[14] Perhaps in compensation many New Zealand plants with small sometimes wind-pollinated flowers (Fig. 5) produce an abundance of brightly coloured berries which are bird dispersed. Paradoxically, a number of the plants of the New Zealand subantarctic islands have flowers more brightly coloured than their New Zealand relatives even though these islands too lack specialised pollinators.[11]

*Figure 5* Kanono (*Coprosma grandifolia*): A New Zealand example of both wind pollination and dioecism. The flowers are male. As is generally the case with wind-pollinated flowers they are small and inconspicuous, but have disproportionately large dangling stamens with large anthers. The hanging stamens move readily with the wind, which shakes out the large quantities of pollen necessary for this rather wasteful method of pollination. The family to which *Coprosma* belongs, the Rubiaceae, is mostly insect-pollinated with often showy flowers. The species of *Coprosma* are unusual in the family in being both dioecious and wind-pollinated.
Photo: M. D. King.

Probably the isolated island characteristics in the New Zealand flora developed after isolation had been attained, and, more particularly, since high mountains and glacial/interglacial fluctuations have developed in recent geological times. When mountains first became significant and temperatures cooled sufficiently the climate would have become too cold on the mountain tops for tree growth and alpine habitats would have come into existence like new islands, although in this case not surrounded by a sea of water, but by a sea of mostly unsuitable plants. Because of New Zealand's isolation, plants suited to cold conditions elsewhere could arrive only slowly by long distance dispersal so there would have been plenty of opportunity for the establishment of hardy variants of forest genera in progressively colder habitats. It was probably in these circumstances that genera such as *Coprosma* attained their present, unusually wide, morphological and ecological ranges by radiating into non-forest habitats.

As each glacial phase progressed the area of alpine vegetation would extend, reaching sea level in many places. The forests would become greatly restricted with some species becoming extinct and others barely surviving. With the warmer conditions of the interglacials the situation would be reversed and the alpine flora would come under pressure. Such climatic fluctuations, combined with an isolation that would prevent both the ready recruitment by long distance dispersal of species into new habitats, and the ready survival by migration of species already present, would put variable populations, including dioecious species and hybrid swarms, at a distinct advantage. In times favourable to their vegetation type they would be best able to take advantage of new habitats and in unfavourable times they would be more likely to survive.

## Alternative Views

The preceding analysis concludes that both long distance dispersal and former land connections have played a role in the ancestry of the New Zealand flora. This would, I think, be the majority view at the present time, but there are alternatives: on the one hand Mildenhall[15] considers the possibility of long distance dispersal even for *Nothofagus*, while on the other the 'panbiogeographers' do not think speculation on dispersal has any value when trying to explain the presence of certain taxa on isolated islands.[16]

Alternatives to long distance dispersal as an explanation for the floras of isolated islands include the suggestion of Melville[17] and others that there was formerly a continent, Pacifica, which rifted into a number of fragments that drifted through the Pacific and eventually became incorporated into various continents on the Pacific rim. If true, this would mean that the isolated central Pacific islands or their predecessors may have been close to continental fragments in the past and could have derived the nuclei of the present island floras from them. It is further suggested that Pacifica was originally sited in the south-west Pacific and that it may have provided the sediments that now form the rocks of eastern New Zealand.

Carey[18] has promoted a different idea for some time — that the earth is expanding. When it was much smaller all the present land areas were together without intervening oceans. As the earth expanded the land areas moved apart and the new basins between them became filled with water derived from the earth's interior. In this case too the islands of the young small Pacific Ocean could have derived their floras from the then nearby continents.

## Other Special Features of the Flora

Other striking and puzzling features of the New Zealand flora will be considered in later chapters. These include:

(a) the abundance of specialised growth forms in the conifer broad-leaf forests that are similar to growth forms of tropical forests;

(b) the prevalence of distinct and varied juvenile forms in a number of the forest species;

(c) the local abundance of very small-leaved, densely twiggy, springy, often cushion-like shrubs (divaricating shrubs) belonging to many species and ranging from forest and open lowland habitats to beyond tree line;

(d) specialised alpine growth forms such as cushion plants and scree plants.

## CHAPTER 2

# Conifer Broadleaf Forest: General Features and Tropical Comparisons

*Ferns grow everywhere, clinging like ivy to the rough stems, festooning them with elegant fronds, webbing them with veils of delicate rhizome, overrunning fallen boughs, drooping long languorous growths from matted clumps overhead. Rooted in massy forks grow epiphytes such as* Griselinia lucida *and huge rookeries of pineapple-like* Astelia. *Mats of sweet-scented orchids cling with a plexus of roots to suitable sites. There is a luxuriance of growth due to the great rainfall and the large number of hours of sunshine, almost unknown elsewhere. The edges of the forest exhibit a still more voluptuous profusion of tangled growth — clematis, rubus, vine, parsonsia, and the native passion-flower competing in the ampler light. Such a forest as this, typical of the North Island, is in truth tropical in all except degree, in all except latitude . . . an exuberance of life prevails, a luxuriance unknown elsewhere save in the true tropical zone.*

This quotation from Guthrie-Smith's *Tutira*[19] tends to purple Victorian prose perhaps, but is typical of the reaction of European settlers in New Zealand to the lowland forest or 'bush' (Figs. 6, 7, 8), so unlike the forests of 'home'. A more professional but similar reaction comes in more recent times from a forester from the tropics.[20]

*I was astonished to find on my arrival that one of the main types of indigenous forest is very similar to Malayan rain forest; far closer to the latter in its structure and appearance than to any type of European forest.*

New Zealanders who have grown up with the 'bush' do not see it as unusual. The reason why visitors from comparable latitudes in the northern hemisphere are so taken by surprise is that they expect to find forests similar to those *they* grew up with: the mid-latitude deciduous forests and the higher latitude coniferous forests with their relatively few species, simple structure and general lack of specialised vines and epiphytes. Instead they find in the lowlands of New Zealand, particularly at lower northern latitudes, a forest which apparently exhibits all the features they had associated with, and perhaps observed

*Figure 6 (opposite)* Conifer broadleaf forest. Inland Taranaki, west central North Island.
Photo: J. W. Dawson.

*Figure 7* Conifer broadleaf forest south of Kaitaia, northern North Island. The emergent trees on the ridge crest are mostly rimus (*Dacrydium cupressinum*). Photo: B. V. Sneddon.

en route, in forests of the tropics. This being the case, in the following review, consideration of each of the features of mostly free standing plants of New Zealand's conifer broadleaf forest will be preceded by an outline of the comparable features in tropical rain forest.[21] Forest plants which depend on trees for mechanical support (vines and epiphytes) or nutriment (parasites) will be considered in the next chapter.

## Numbers of Species

Of the major types of world vegetation, tropical rain forest is the most complex in structure and probably the richest in species. It is found

*Figure 8* Interior view of conifer broadleaf forest south of Kaitaia, northern North Island. In the foreground are young plants of the nikau palm (*Rhopalostylis sapida*) and the shrub kawakawa (*Macropiper excelsum*). A larger nikau is on the right. In the middle distance the shrubs are mostly young plants of the subcanopy tree kohekohe (*Dysoxylum spectabile*). The trunks to the left belong to the canopy dominant taraire (*Beilschmiedia tarairi*) and the large, partly obscured trunk at the centre to the emergent northern rata (*Metrosideros robusta*). Photo: B. V. Sneddon.

mostly between the tropics of Capricorn and Cancer in regions where rainfall is abundant and evenly spread throughout the year. The three major regions are west central Africa; south-east Asia to the Pacific; and northern South America and Central America. These three regions are widely separated geographically and although the forests are vegetationally very similar, each has its own species and in some cases its own genera. The total numbers of species in these forests are impressive. Tree species alone are often numbered in the hundreds at many localities, compared with the dozens found in temperate forests. In most lowland tropical forests the trees are all flowering plants.

In the New Zealand conifer broadleaf forest, tree species are numbered in dozens rather than the hundreds found in tropical rain forest. Some see this as a major difference precluding any suggestion of a close relationship between the two types of forest. However, this conclusion may not be justified — species richness is not the only distinctive feature of tropical rain forest as we shall see.

The conifer broadleaf forest, as the name implies, comprises a mixture of conifers and flowering plants.

## Stratification

Structurally, the tropical rain forest is considered to have five strata; three tree layers, a layer of shrubs and a ground layer of herbaceous plants. This contrasts with most temperate forests where, at most, three strata are recognised; tree, shrub and ground.

The uppermost layer of very tall trees in the tropical rain forest is often discontinuous and the individual trees are referred to as emergents. This might be taken to imply that the 'emergents' grow through and above the canopy formed by the second tree layer, but studies indicate that many of them are light-demanding species which establish themselves early in a forest's history, grow rapidly, and initially form a continuous layer. As the lower forest layers close up, the forest floor becomes more shaded, many emergents cease to regenerate, and in time the highest tree stratum becomes discontinuous as old emergents die and are replaced only sporadically in canopy gaps.

In the conifer broadleaf forest five strata can usually be recognised, although they are lower in stature than their tropical equivalents:

(a) Emergent trees (30–40 m): mostly conifers.
(b) Canopy trees (20–25 m).
(c) Subcanopy trees (10–15 m): including our only native palm, the nikau *(Rhopalostylis sapida)* and *Cyathea* and *Dicksonia* tree ferns.
(d) Shrubs and small trees (3–8 m).
(e) Ground plants (0–1 m): mostly ferns, but some flowering plants particularly in the north.

## Specialised Roots

A number of tropical trees exhibit structural features which appear strange to the eyes of those from most temperate regions. Many trees, particularly in swampy situations, develop thin flanges or plank buttresses from the bases of their trunks. It is suggested that these plank buttresses, which are more or less triangular in form and may extend for several metres up the trunks and out along the roots, confer greater stability on the often shallowly rooted trees. These plank buttresses are very inconvenient for tree fellers who often find it necessary to construct scaffolding so that the trunk can be sawn through above them.

Other trees increase support for their trunks by forming downward-arching prop roots. This can be seen, perhaps most conspicuously, in mangrove forests. In swampy situations many trees send up pneumatophores or breathing roots, through which air is taken in and transferred to the root system beneath the swamp. Pneumatophores are particularly conspicuous in saline coastal swamps inhabited by mangrove forests, but are also found in inland freshwater swamps. An unusual type of aerial root in tropical forests is the column root, which descends to the ground from horizontal branches, such as in the banyan figs. Individual banyan trees can sometimes spread out by this means over hectares of ground.

### Plank Buttresses

In New Zealand pukatea (*Laurelia novae-zelandiae*) often has well defined plank buttresses (Fig. 9). Other trees may also have buttresses: e.g. kohekohe (*Dysoxylum spectabile*), kahikatea (*Dacrycarpus dacrydioides*), and some species of beech (*Nothofagus*), but these are usually not plank-like.

### Pneumatophores

Pukatea, when growing in swamps, also forms large pneumatophores. These are basically shield-shaped and sometimes several times higher than they are wide (Figs. 10, 11). These pneumatophores originate when a root tip arches above the swamp surface and then grows back in again so forming a loop. Wood is then added mostly on the upper and lower sides of the loop which leads to the eventual shield-shape. The pneumatophores are covered with mosses and have pustules (lenticels) of loose corky cells, through which air enters the root system.

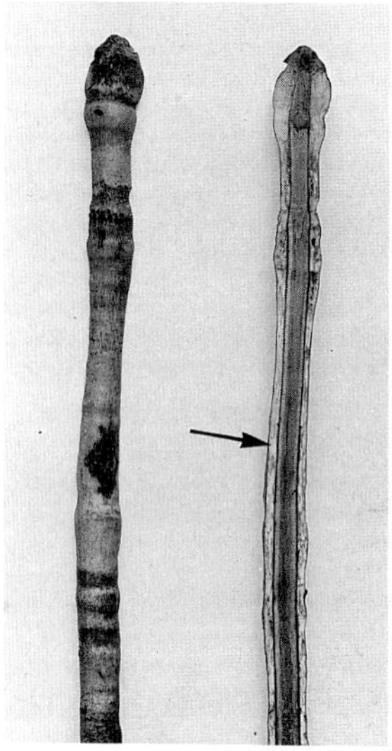

Swamp maire (*Syzygium maire*) is often associated with pukatea in swamp forests and it too forms specialised pneumatophores. In this case, however, they are not woody loops, but smaller, finger-like and rather spongy root tips several centimetres long. They are orangey-brown in colour and often branch near the base to form coral-like clusters. A longitudinal section through such a 'peg root' reveals a white outer cylinder of air-filled tissue (Fig. 12).

In the same swamps, kiekie (*Freycinetia baueriana* var. *banksii*) often sprawls on the forest floor as well as climbing up tree trunks. In the former situation it forms what may be pneumatophores of an unusual type. They are slender and finger-like, and have a succession of tyre-like rings of white aeration tissue (Fig. 10, at lower left).

Finally the pneumatophores which catch the eye of many people in the Auckland region are those formed by the mangroves. These too are finger-like and project above the mud surface at low tide (Fig. 13).

*Figure 9 (above right)* Plank buttresses of pukatea (*Laurelia novae-zelandiae*). Photo: F. B. Sampson.

*Figure 10 (above far right)* Pneumatophore of pukatea (*Laurelia novae-zelandiae*). Photo: M. D. King.

*Figure 11 (below right)* Pneumatophore of pukatea (*Laurelia novae-zelandiae*) showing the junction between the two sides of the original loop root. Nga Manu Reserve, Waikanae, southern North Island. Photo: J. E. Casey.

*Figure 12 (above)* Pneumatophores of swamp maire (*Syzygium maire*). The one on the right has been cut in half longitudinally. The arrow indicates the white air-filled tissue. Photo: J. E. Casey.

*Figure 13 (below far right)* Pneumatophores of mangrove (*Avicennia resinifera*). Near Kaeo, northern North Island. Photo: J. W. Dawson.

### Prop Roots

In the swamp maire the primary root in young plants often atrophies or becomes weak, and prop roots develop from the base of the trunk to provide the root system of the adult tree. In young trees the space between the base of the trunk and the ground is readily observable (Fig. 14).

The nikau palm may also produce a fringe of slender prop roots a little above ground level.

### Column Roots

Pohutukawa (*Metrosideros excelsa*) forms aerial roots very readily, which run down the trunk to the ground. Sometimes it forms column roots from more-or-less horizontal branches. Except where the trees have a semi-sprawling habit on coastal cliffs, these roots often do not reach the ground, but branch profusely at the lower end to give a straw broom effect (Fig. 15). Similar roots can be observed in some tropical figs (*Ficus*).

## Cauliflory

A strange habit of some tropical trees is the production of flowers directly on the trunks and/or branches, a phenomenon known as cauliflory. The most notable New Zealand example of this is kohekohe

*Figure 14 (above left)* Roots of swamp maire (*Syzygium maire*). The arrow indicates the space below the base of the trunk where the primary root failed to develop. Nga Manu Reserve, Waikanae, near Wellington, southern North Island.
Photo: J. E. Casey.

*Figure 15 (below far left)* Aerial roots of pohutukawa (*Metrosideros excelsa*).
Photo: J. W. Dawson.

*Figure 16 (below left)* Inflorescence arising directly from the trunk of kohekohe (*Dysoxylum spectabile*). Photo: M. D. King.

*Figure 17* Ramiflory. Female flowers of mahoe (*Melicytus ramiflorus*) borne on woody twigs.
Photo: M. D. King.

(*Dysoxylum spectabile*) where the sprays of orange-blossom-like flowers may be produced directly on the trunks (Fig. 16) as well as on the major branches.

The flowers of the tree fuchsia (*Fuchsia excorticata*) are mostly produced towards the ends of woody branches, but sometimes appear directly on the trunk.

Two rare species on the Three Kings Islands, *Tecomanthe speciosa* and *Pennantia baylisiana* are also cauliflorous.

In the species of *Myrsine*, *Melicytus* (Fig. 17) the liane *Metrosideros diffusa* and *M. parkinsonii*, the flowers arise on woody twigs towards the tips of branches, a less extreme condition often known as ramiflory.

## Leaf Features

The leaves of plants of moist tropical forests are evergreen, generally larger than those of temperate forests and are often leathery and smooth-margined. Average leaf size decreases with increasing height above the ground and some of the trees have distinct juvenile forms with leaves much larger and/or more compound than those of the adults. In some cases there are narrow prolongations from the ends of the leaves known as 'drip tips', thought by some to enable rapid drying of leaves after rain.

Pulvini or elastic swellings at one or both ends of the leaf stalk or petiole are a common feature of tropical forest plants. It has been suggested that bending movements at these pulvini enable the leaf to maintain the best orientation to the light.

### Leaf size

The average leaf size is considerably less in the New Zealand conifer broadleaf than in tropical rain forest. This can be demonstrated by comparing the species of a far northern New Zealand forest with a Philippines forest in terms of a now widely accepted leaf size classification.

| *Philippines Forest*[21] | |
|---|---|
| Species with macrophylls (leaves 18-60 cm long*) | 10% |
| Species with mesophylls (leaves 6-18 cm long) | 86% |
| Species with microphylls (leaves 2-6 cm long) | 4% |

| *New Zealand Forest*[22] | |
|---|---|
| Species with macrophylls | 1% |
| Species with mesophylls | 25% |
| Species with microphylls | 68% |
| Species with nanophylls (leaves 1-2 cm long) | 6% |

* Based on leaves about twice as long as they are wide

The smaller leaf sizes in New Zealand probably correlate with lower temperatures.[23]

### Teeth, Pulvini and Drip Tips

The proportion of woody dicotyledon species in the conifer broadleaf forest with smooth-margined leaves or leaflets is lower than that of tropical forests: about 56 per cent compared with 80 per cent recorded in a Nigerian forest.[21] Godley[24] has observed that some New Zealand trees have much more prominent teeth on the leaves of young plants than on those of adults; for example wineberry (*Aristotelia serrata*) and ngaio (*Myoporum laetum*). In other cases, some trees with smooth-margined adult leaves or leaflets have juvenile leaves with toothed or lobed margins; for example puriri (*Vitex lucens*), titoki (*Alectryon excelsus*), kohekohe (*Dysoxylum spectabile*) and several species of *Hebe*.

Pulvini also are not common. The species of maire (*Nestegis*) have dark-coloured pulvini at the petiole bases and kohekohe (*Dysoxylum spectabile*), titoki (*Alectryon excelsus*) and king fern (*Marattia salicina*) have them at the bases of the leaflet petioles. Hinau (*Elaeocarpus dentatus*) is an interesting case. The adult leaves do not have pulvini, but on juvenile leaves they can be observed at each end of the petioles. The larger-leaved tropical species of *Elaeocarpus* have prominent pulvini in the same position on adult leaves.

Drip tips are neither strongly developed nor very common in the New Zealand forest. In fact the most that can be said is that ten or so species tend to have slightly to moderately drawn out leaf tips especially when growing under sheltered, shady conditions. Perhaps both drip tips and pulvini were better developed in the ancestors of our present trees.

## Juvenile Forms[24,25]

Often very distinct juvenile and adult forms are a noticeable and, to those learning to identify native plants, a tiresome feature of the New Zealand flora. In the conifer broadleaf forest, some of the juveniles contrast with those of the tropics in that they are very freely branched with leaves much smaller than on the adults. These will be considered in Chapter 6 as part of the wider question of the small-leaved, freely branched or 'divaricating' shrubs, so prevalent in New Zealand.

The remaining rain forest species under this heading have juvenile leaves which are larger or longer or more compound or more dissected than those of the adults.

Raukawa (*Pseudopanax edgerleyii*) has palmately compound juvenile leaves with up to five deeply lobed leaflets. The adult leaves are simple and smooth-margined.

One variety of *Pseudopanax simplex* has juvenile leaves very similar to those of *P. edgerleyii* and simple, toothed adult leaves (Fig. 18).

Pate (*Schefflera digitata*) is unusual in that juvenile leaves are only found on some plants in the northern half of the North Island. These leaves are palmately compound and similar in size to those of the adults, but the leaflets are deeply lobed rather than just toothed.

In three tree species, all belonging to the family Cunoniaceae, the leaves are basically pinnately compound. In *Ackama rosifolia* the juveniles have six to ten pairs of leaflets, the adults three to five. In towai (*Weinmannia silvicola*) the trend is from five pairs to one pair of leaflets; in kamahi (*W. racemosa*) from one or two pairs of leaflets to simple leaves (Fig. 19).

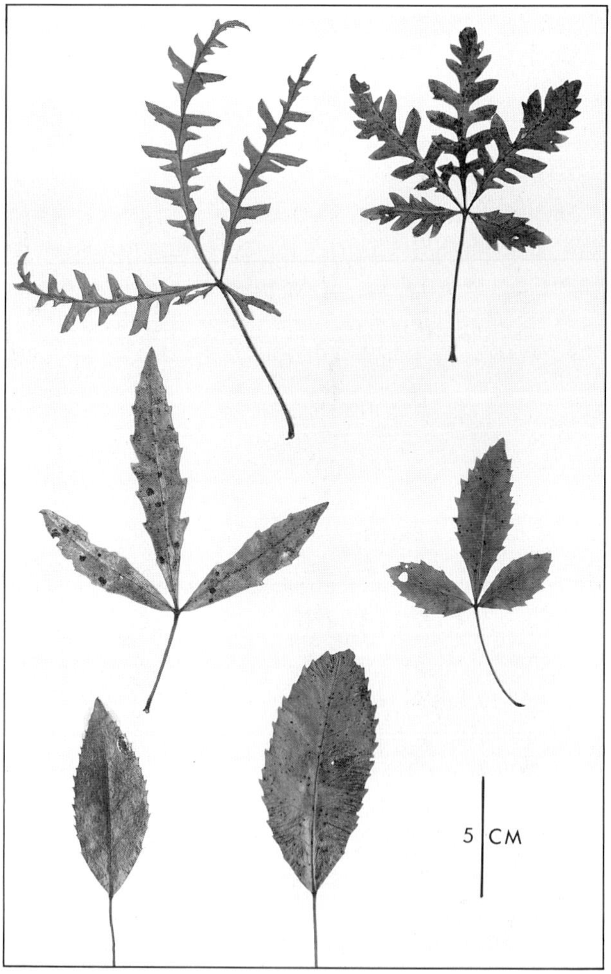


*Figure 18* Juvenile and adult leaves of *Pseudopanax simplex* var *simplex*. Dissected compound juvenile leaves above; serrate, compound semi-juvenile leaves centre; simple, serrate adult leaves below. Photo: J. E. Casey.

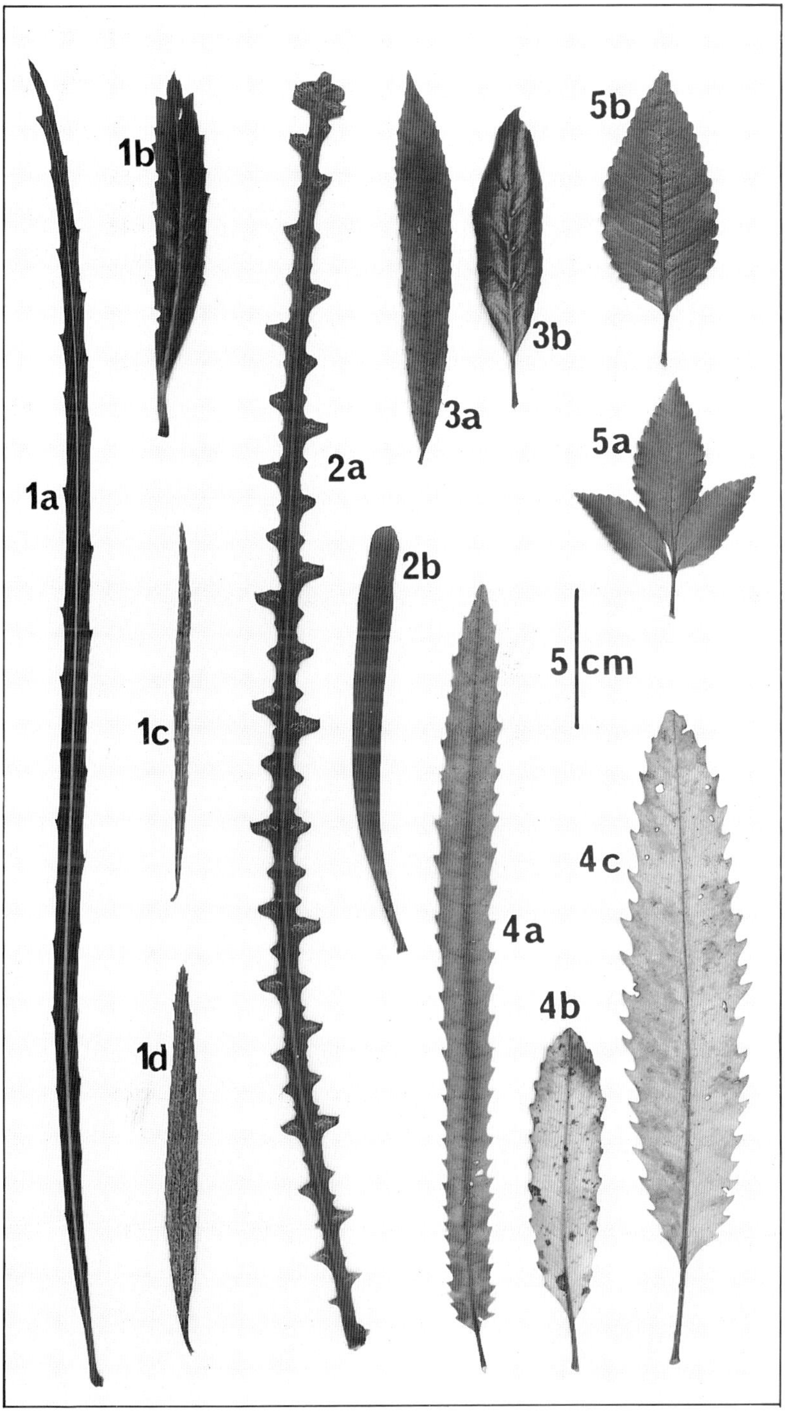


*Figure 19* Juvenile and adult leaves.
Lancewood (*Pseudopanax crassifolius*): 1a, juvenile leaf; 1b, adult leaf; 1c, d, brown, mottled leaves from a seedling.
*Pseudopanax ferox*: 2a, coarsely toothed juvenile leaf; 2b, adult leaf.
Hinau (*Elaeocarpus dentatus*): 3a, juvenile leaf; 3b, adult leaf (note domatia).
Rewarewa (*Knightia excelsa*): 4a, juvenile leaf; 4b, c, adult leaves.
Kamahi (*Weinmannia racemosa*): 5a, trifoliolate juvenile leaf; 5b, simple adult leaf.
Photo J. E. Casey.

In all cases where the juvenile leaves are compound and the adult leaves simple, the latter have a distinct joint at the end of the petiole as an indication of their compound derivation.

Only a few New Zealand species have juvenile leaves which are broad and much larger than the adults: wineberry (*Aristotelia serrata*), *Nestegis*

*Figure 20* Adult lancewood (*Pseudopanax crassifolius*), with a basal shoot of completely juvenile form to the left. Nga Manu Reserve, Waikanae, southern North Island. Photo: J. E. Casey.

*apetala* and *Coprosma chathamica*. In several species of *Dracophyllum*, where all the leaves are long and narrow, the juvenile leaves are much larger than the adults. In species of other genera the juvenile leaves are themselves very small but they are still larger than the adult leaves, which are reduced to scales, as in several species of 'whipcord' *Hebe*, *Helichrysum coralloides* and a group of conifers formerly included in *Dacrydium*: silver pine (*Lagarostrobos colensoi*), yellow silver pine (*Lepidothamnus intermedius*), bog pine (*Halocarpus bidwillii*), *H. biformis* and manoao (*H. kirkii*). In the latter cases there is an abrupt change from juvenile to adult leaves. Two other conifers with scale-leaved adults are our only *Dacrydium* — rimu (*D. cupressinum*) (Fig. 21) and kahikatea (*Dacrycarpus dacrydioides*) (Fig. 22). In both of these the juvenile leaves are small and needle-like. The juveniles of rimu are notable for their attractive weeping habit.

*Figure 21 (left)* Juvenile (below) and adult foliage of rimu (*Dacrydium cupressinum*). Photo: J. E. Casey.

*Figure 22 (right)* Juvenile (above) and adult foliage of kahikatea (*Dacrycarpus dacrydioides*). Photo: J. E. Casey.

In most of the species that follow, juvenile leaves are as narrow as the adult or narrower, but may have a larger area by virtue of their greater length. This is not a pattern familiar in the tropics, although on Mauritius and Reunion Islands in the Indian Ocean there is a partly comparable pattern — one species each of 24 genera has juvenile leaves which are much narrower than the adult leaves and also much smaller in area.[26]

Lancewood (*Pseudopanax crassifolius*) (Figs. 19, 20) is the best known of our trees with a juvenile form.[27] The juvenile leaves are several times longer than those of the adult and a little narrower. They hang down in a distinctive cluster from the tip of the slender stem, which does not branch until it attains a height of 4–5 m after 15 or more years. The stem is then able to branch into a dense and rounded crown with short, broad, upwardly directed adult leaves. The related but less common *P. ferox* is very similar, but the juvenile leaves have very coarse irregularly-shaped teeth (Fig. 19). Lancewood also has a distinct seedling form in which the leaves are very variable in shape, size and degree of dissection.

Rewarewa (*Knightia excelsa*) and hinau (*Elaeocarpus dentatus*) (Fig. 19) with their long narrow juvenile leaves have a juvenile-adult pattern similar to lancewood, but branching is initiated at an earlier stage. The juvenile leaves of hinau are soft, widest near the tip and have obscure teeth; those of rewarewa are stiff, of even width and have coarse teeth. As mentioned earlier, the juvenile leaves of hinau may have pulvini, which the adult does not, but the adult leaves have pouch-like cavities or domatia (Fig. 23),[28] where the secondary veins meet the midrib on the undersides.

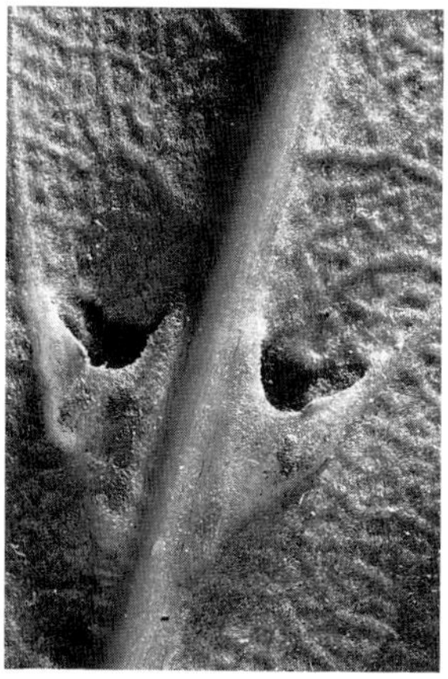

*Figure 23* Domatia of the adult leaves of hinau (*Elaeocarpus dentatus*). The pouch-like domatia form at the junctions of the secondary veins and the midrib. Photo: M. D. King.

Three of the four species of *Nestegis* or 'native olives' have narrow adult leaves. In the case of two species, black maire (*N. cunninghamii*) and white maire (*N. lanceolata*), the juveniles are much narrower than the adults, but as they are also often longer they may have about the same area as the adult leaves or somewhat less. The last species, *N. montana*, has adult and juvenile leaves which are almost equally narrow, but the juveniles, being longer, are greater in area than the adults.

Trees with distinct juveniles may exhibit a curious phenomenon whereby fully adult specimens give rise to shoots near the ground which bear leaves of juvenile form (Fig. 20). These are termed reversion shoots and they derive from dormant buds formed during the juvenile phase of the tree and so are 'programmed' to be juvenile. Pokaka (*Elaeocarpus *hookerianus*) may be an exception, as reversion shoots can occur 10 m or more above the ground; well above the height of the juvenile stage.

### The Deciduous Habit

In north temperate forests many trees and shrubs are leafless during the winter, and in tropical forests some species, growing in areas where there is a distinct dry season, may be leafless during the unfavourable period. In the New Zealand rain forest, which grows in parts of the country where there is no well-marked dry season and winters are relatively mild, the deciduous habit is uncommon.[29,30,31] The following New Zealand conifer broadleaf forest species are or may be leafless during winter: wineberry (*Aristotelia serrata*), *Fuchsia excorticata,* the vines *Muehlenbeckia australis* and *M. complexa,* and ribbonwood (*Plagianthus regius*). Some forms of kowhai (*Sophora microphylla*) lose their leaves in the spring before flowering. However, with at least *Fuchsia* and *Aristotelia,* leaf fall is more related to temperature than to declining day length, which is the triggering factor for most northern temperate deciduous trees. Both species may be evergreen in milder latitudes.[63]

## Bud Protection

Unlike many trees of temperate climates, where the immature leaves at the branch tips are protected during the winter by dry overlapping scales, the buds of many tropical trees of moist climates are not pro-

*Figure 24* Tarata or lemonwood (*Pittosporum eugenioides*). Overwintering buds with overlapping bud scales. Photo: M. D. King.

tected in this way. This is understandable as they are not subject to an unfavourable season. Nevertheless the delicate immature leaves are at some risk of drying out and they may be protected by a layer of hairs or by mucilaginous or resinous secretions. The petiole bases of mature leaves adjacent to a bud may also afford some protection by partly enclosing it with or without the assistance of paired appendages known as stipules.

In this respect the woody flowering plants of the New Zealand conifer broadleaf forest seem comparable to those of the tropics. Of 45 genera only 3 have numerous, relatively large bud scales — *Pittosporum* (Fig. 24), *Aristotelia* and the tree species of *Metrosideros*. In the beech forest (see Chapter 5) the species of *Nothofagus* also have many overlapping bud scales.

Nineteen genera have no bud protective structures at all although the immature leaves are often hairy (Fig. 25). Admittedly, genera in this group with strong tropical affinities, for example *Dysoxylum, Vitex, Beilschmiedia, Litsea,* are represented by only one or two species in New Zealand.

The remaining genera afford their immature leaves partial or complete protection with a few small scales or with the stipules or sheathing bases of adjacent mature or developing leaves. Some also form protective secretions such as mucilage in *Coprosma* and a varnish-like material in some species of *Pseudopanax*.

*Figure 25* Rewarewa (*Knightia excelsa*). Very young leaves with a furry pubescence, but otherwise unprotected. Photo: M. D. King.

CHAPTER 3

# Conifer Broadleaf Forest: Vines, Epiphytes and Parasites

Woody and herbaceous vines, vascular epiphytes or 'perching plants', and vascular parasites and saprophytes are abundant and distinctive features of tropical rain forests.[21] The stems of the vines climb their supports by a variety of means, including twining of the stems, tendrils, hooks and attaching roots. The epiphytes range from herbaceous plants to large trees. Among the former are nest epiphytes, with their large leaf rosettes, and an abundance of orchids, and among the latter are the 'strangling' epiphytes, including many species of fig (*Ficus*). The parasites attack stems or roots and may be without chlorophyll and completely parasitic or with chlorophyll and so able to manufacture some of their food requirements. Saprophytes are devoid of chlorophyll and are thought to derive their energy requirements from decaying organic material, particularly leaf litter, on the forest floor.

Surprisingly, in view of our temperate latitudes, a comparable range of vines and epiphytes are equally conspicuous in the New Zealand conifer broadleaf forest, but these have received little attention in books on New Zealand forest plants for the general reader. This partial neglect is perhaps understandable, as it is not easy to get close to the foliage, flowers and fruits of many of these plants, particularly where they occur high in the branches of tall trees. Binoculars help, but few botanists are as intrepid as J. L. Harrison-Smith,[32] who spent many hours 'wandering about' among the branches of giant kauris recording epiphytes. In his own words: 'In order to catalogue the plants the trees were climbed. Ordinary gum collectors' climbing gear (spiked boots and climbing hooks) were used. Descent was accomplished with a "bo'sun's chair" suspended from a rope. It was then possible to examine the trunk and any isolated branches on the way down.' Fortunately it is possible to reach most vines and epiphytes without taking such risks. Some species grow low on the trunks of trees and so are readily accessible, but even those normally restricted to tree crowns may descend to lower levels in well-lit situations and, in the normal course of events, trees are blown or fall over (Fig. 26) and then a full range of high epiphytes and vines, albeit somewhat damaged, can be examined.

Not infrequently the categories of vines, epiphytes and parasites are confused, so that if it is stated that a plant 'grows epiphytically on trees', further enquiry is often necessary to determine the precise mode of growth of the species in question. It is true that the word 'epiphyte', meaning 'a plant that grows on other plants', fits all three categories, but as in other respects their life styles are quite distinct, this term should be restricted to plants which germinate and establish themselves on the trunk and branch surfaces of trees. Vines, by contrast, establish themselves on the ground, then grow up the trees. Some parasites are similar to epiphytes in that their seeds germinate on trunks and branches, but they differ markedly in having special root-like organs that penetrate the living tissues of their host and draw water and nutrients from them.

## Vines

### Subcanopy Climbers

Vines in this category are herbaceous and attach themselves to tree trunks by special roots arising from the stems. They ascend for varying distances up the trunks, but mostly do not enter the tree crowns. They are all able to reproduce in the reduced light of the forest interior.

In New Zealand all vines in this category are ferns and in some, as in the climbing members of the arum lily family in the tropics, there is a remarkable increase in size and complexity of the leaves as their height above the ground increases. The best known example of this is *Blechnum filiforme* (Fig. 27), a common plant in lowland forest as far south as the northern South Island. *Blechnum filiforme* is often abundant on the forest floor where it spreads by slender rhizomes.[33] In this situation the leaves are only about 10 cm long and once-pinnate, with leaflets which range in shape from small and oblong to almost round. Where the stems grow up tree trunks, as they frequently do, the leaves produced become progressively larger, until several metres above the ground they attain a maximum length of almost 35 cm and have long,

*Figure 26* Kahikatea (*Dacrycarpus dacrydioides*) with asteliad nest epiphytes. The tree originally had two trunks. The nearest one has fallen and its crown and epiphytes can be seen in the foreground. Te Marua near Wellington, southern North Island. Photo: M. D. King.

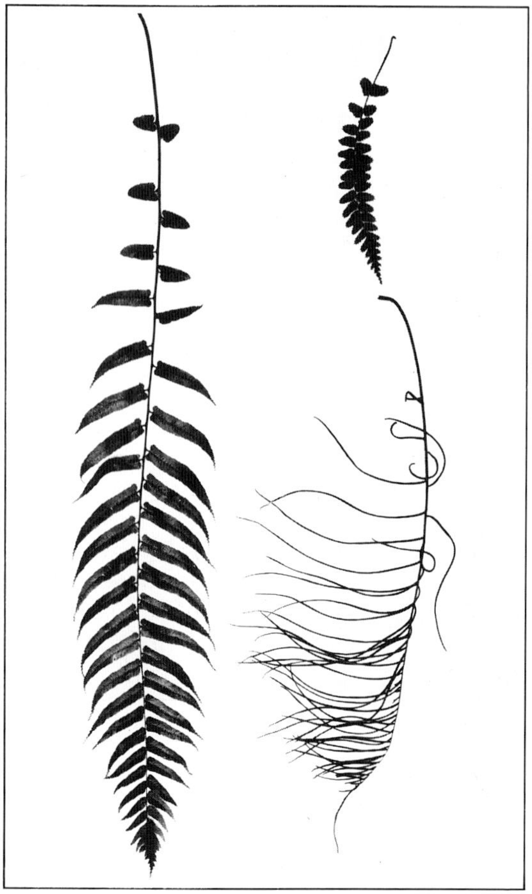

narrow, pointed leaflets up to 10 cm long. It is among these largest leaves that the fertile, spore-bearing fronds, with their almost thread-like leaflets, are produced. The leaves of this species are dark-green and fairly thin. The climbing stems branch and the branches tend to stay close together and grow more or less vertically, sometimes reaching as high as 10 m above the ground.

*Arthropteris tenella* (Fig. 28) also has relatively thin, dark leaves and slender, vertically ascending stems, but these generally reach to only a few metres above the ground. This species also reaches the northern South Island, but is less common than *Blechnum filiforme* and is mostly encountered in coastal forests and some low altitude inland locations. The juvenile leaves are 5 10 cm long, once-pinnate and often rather peculiar in appearance — the few pairs of lateral leaflets are small and

*Figure 27 (left)* Leaves of the climbing fern *Blechnum filiforme*. Top right, a small leaf from the base of a tree trunk. Left, large leaf from 2 m above ground level. Lower right, fertile leaf. Photo: M. D. King.

*Figure 28 (right)* Leaves of the climbing fern *Arthropteris tenella*. Top left, climbing stem with small juvenile leaves. Top right, large adult leaf. The other leaves are at intermediate stages. Photo J. E. Casey.

*Figure 29 Phymatosorus scandens* growing up a tree trunk. The simple leaves are juvenile and compound leaves adult. Photo: J. E. Casey.

almost circular; the terminal leaflet is much larger and longer, narrowing to a point. Fully adult leaves are 30 cm or more long with many pairs of narrow, wavy-margined leaflets up to 8 cm long.

The light green, thin-leaved *Phymatosorus scandens* also shows a trend from juvenile to adult leaves with increasing height (Fig. 29), but in this case the juveniles are about as long as the adults, up to 35 cm, but are narrow, undivided and usually sterile. There is often a fairly abrupt change to the adult leaves, which are much wider and are deeply incised into a number of narrow, lateral segments bearing sporangia.[34] This species often occurs with *Blechnum filiforme* in lowland forests, sometimes on the same tree, but it ranges further to the south of the South Island. It is able to climb vertical tree trunks with its slender stems spreading in various directions, but it does not often attain the heights of *Blechnum filiforme*.

The thick-leaved *Phymatosorus diversifolius,* with its stout, grey-green, black-flecked stems, is better known to most people. It ascends to higher altitudes than the species so far considered and also reaches the Auckland Islands to the south of New Zealand. It differs too in often preferring to climb inclined trunks and inclined or horizontal branches, so it is most common on trees such as mahoe, tree *Fuchsia* and kamahi, all of which have short trunks and many spreading branches. This fern may extend along all the branches and eventually into the crown of such trees. The stems branch freely and often rather untidily on their supports, sometimes curving completely around them. Where *Phymatosorus diversifolius* grows on the upper sides of more or less horizontal branches, quite a thick layer of humus builds up beneath its stems.

As the name of this species indicates, it has a range of leaf forms similar to *P. scandens*. On a tree with a heavy growth of the fern, the large, shiny, bright green leaves are rather widely spaced and deeply incised into narrow segments with an abundance of sporangia beneath, aggregated into distinctive orange spots or sori. Young plants establishing themselves in moss on trunk bases have narrow, undivided sterile leaves. *Phymatosorus diversifolius* also grows on the ground, most abundantly on rocky slopes. In rocky, exposed places the narrow, undivided leaves may persist, but in these circumstances they bear sporangia.

A third species of *Phymatosorus* — *P. novae-zelandiae* — is found in montane forests throughout the North Island, but is absent from the South Island. With its stout rhizomes it is similar to *P. diversifolius,* but the rhizomes are densely covered with straw-coloured scales and the leaves are generally larger with more numerous, narrower and longer lateral segments. It appears that there are no marked variations in leaf form in this species.

*Rumohra adiantiformis* ranges throughout New Zealand in lowland to montane forests and is most common as a climber on tree fern trunks. The much divided leaves have a leathery texture and bear conspicuous black sori. There is a modest increase in the size of leaves with increasing height.

Filmy ferns may be common as climbers in high rainfall areas. In some cases the fronds are very small and delicate and grow intermingled with mosses, but some species — *Hymenophyllum dilatatum, H. scabrum, H. sanguinolentum* and several others — have relatively large leaves which, Holloway[35] notes, increase in size with increasing height above the ground. Holloway suggests that the increased leaf areas enable more effective absorption of water by the thin leaves in the drier tree trunk habitat. The kidney fern (*Trichomanes reniforme*) is perhaps our most unusual filmy fern. Its leaves are undivided and, as both common and botanical names indicate, kidney-shaped. This fern usually grows for only a short distance up tree trunks, but can climb much higher in moist situations. A strange case is the densely hairy *Hymenophyllum malingii,* which climbs on the dead trunks of trees, particularly mountain cedar (*Libocedrus bidwillii*). The climbing species of filmy fern range throughout the country.

Six of the climbing ferns considered here are restricted to New Zealand. The ranges of others are:

*Phymatosorus diversifolius:* Australia, Tasmania, tropical Polynesia.

*P. scandens:* Australia, Norfolk Island.

*Arthropteris tenella:* Australia, Norfolk Island, New Caledonia.

*Rumohra adiantiformis:* South temperate zone, tropical Polynesia, tropical America.

## Canopy Climbers

Vines whose foliage eventually spreads into the canopy of the forest are usually woody and are often referred to as lianes.[36] At the adult stage they are light-demanding and generally produce flowers, or spores, only in well-lit situations. Young plants on the forest floor are more shade-tolerant, but nevertheless establish most abundantly in the better lit earlier stages in forest development or in canopy gaps in mature forest. Unlike the sub-canopy climbers, the lianes climb by a variety of means — attaching roots, twining stems, hooks and tendrils.

## Root Climbers

The most prominent root climbers are the climbing ratas,[37] currently included in the genus *Metrosideros*. They are able to grow up quite large trunks and are perhaps most abundant on the emergent conifers and the northern rata. Their climbing stems are usually quite slender and the leaves, which form a close mosaic on the tree trunk (Fig. 30), are generally smaller, thinner and more rounded than those of the adult stage. When the stems reach full light high in the tree crown, or adequate light in the lower levels, they form a bushy growth of branches which extend away from the support and eventually bear flowers. At this stage the stems extending up from the ground enlarge considerably and swing away from the host trunk as woody cables (Fig. 31). *Metrosideros fulgens* and *M. perforata* form the largest stems, sometimes up to 15 cm or more in diameter, but the others may attain 7–8 cm. Often no leaves are visible near the ground, but the stems can be identified to some extent from the bark — *M. perforata* has red-brown stringy bark, *M. fulgens* also red-brown bark separating in thickish strips and the other species have pale whitish bark separating in thin flakes.

*Metrosideros albiflora* and *M. carminea* are restricted to the northern North Island. *M. albiflora* is most common in kauri forests. It has the largest leaves of the group and small, white flowers. *M. carminea* is much more colourful, with masses of large, crimson flowers, and it is now popular as a garden plant. *M. perforata, M. fulgens* and *M. diffusa* frequently occur together as far as the northern South Island and *M. diffusa* continues alone to Stewart Island. *M. diffusa* often forms slender stems near the ground which spread widely in the humus of the forest floor and climb any trunks they encounter. *M. colensoi* also reaches the northern South Island, but is more localised in its occurrence, favouring forests on fertile soils such as those of river terraces. The strongly weeping habit of its foliage is a distinctive feature. *M. fulgens* has large flowers ranging in colour from orange to dark red, while the other species have small white to pinkish flowers. Most of the climbing ratas flower from early to mid-summer, but *M. carminea* flowers in early spring and *M. fulgens* is remarkable both for the timing and length of its flowering season — the first flowers may appear in late summer and flowering continues through the winter into early spring.

New Zealand is not the only place where climbing species of *Metro-*

*Figure 30 (above)* Young stage of white climbing rata (*Metrosideros perforata*) forming a leaf mosaic on a tree trunk. Photo: B. V. Sneddon.

*Figure 31 (left)* The cable-like stems of a mature *Metrosideros perforata* on a rimu (*Dacrydium cupressinum*). Kaitoke, near Wellington, southern North Island. Photo: M. D. King.

*Figure 32* The drooping stems and foliage of kiekie (*Freycinetia baueriana* var. *banksii*) completely obscuring a kahikatea trunk. Kaitoke, near Wellington, southern North Island.
Photo: J. W. Dawson.

*sideros* occur. There are climbers related to ours in New Guinea and the Philippines.

The only other root climbing liane in New Zealand is the kiekie (*Freycinetia baueriana* var. *banksii*). It belongs to a distinctly tropical family, the Pandanaceae, which is represented by many species of *Freycinetia* and *Pandanus* in tropical rain forests. One might expect that the outlying New Zealand species would be of reduced form and perhaps rare. In fact it compares with the largest and most robust tropical species and is abundant in lowland, especially swampy forests as far as the south west of the South Island. Tree trunks are often completely obscured by the foliage of kiekie (Fig. 32), which can extend into the highest crowns 30 m or more above the ground. The leaves are dark green, narrow and a metre or more long with finely toothed cutting edges. The male and female inflorescences found on separate plants are cone-like and surrounded by leaf-like white or purplish bracts, which are sweet and edible. The stems are a few centimetres in diameter and distinctively ringed with leaf scars. They give rise to slender, attaching roots, which branch freely towards their ends and attach themselves firmly to the trunk. Other roots are stouter and grow down the trunk to the ground, often building up into quite thick and rather untidy masses.

**Twining Stem Climbers**

Twining lianes have climbing stems which wind around their supports in a clockwise or anticlockwise direction, depending on the species, until they reach the full light of the forest canopy. Unlike root climbers, many twiners are not able to climb large tree trunks — their turning circle is too small for that — so they either have to climb young slender trees and grow with them into the canopy or climb small or young subcanopy trees and transfer from their crowns to those of taller trees. Many twiners also climb stems of their own species which have already gained the forest roof.

Undoubtedly supplejack (*Ripogonum scandens*), which ranges throughout the country, is the most familiar twiner in New Zealand forests,[38] particularly on alluvial and swampy sites. It belongs to the lily family, taken in a wide sense, and its almost black, jointed, bamboo-like climbing stems often form entanglements that greatly impede progress (Fig. 33). Fortunately, unlike several of its relatives in Australia, supplejack does not have prickles. Aggregations of woody, tuber-like rhizomes below ground give rise to the climbing stems which are dark-brown to black, 1–2 cm in diameter and which bear pairs of long narrow, often twisted scales in place of leaves. The stem tips are reminiscent of *Asparagus* and are soft and easily broken. They can elongate at an average rate of 5 cm per day in summer and while growing upwards the upper part of the shoot revolves slowly in an anticlockwise direction. If it does not encounter a support it bends down to the ground and grows up again from the tip.

Supplejack mostly climbs fairly slender supports, but can also twine around quite large trunks, the record being a kohekohe of 1.5 m diameter.[38] When a climbing stem reaches the forest canopy, lateral climbing stems arise from its upper parts and eventually bear relatively slender, leafy stems of limited growth, which are unable to twine. The leaves are broad and distinctively veined with two strong lateral veins more or less parallel to the midrib. The leafy stems bear small flowers followed by bright red berries. When lateral climbing stems are formed near the ground they are often swollen and tuber-like at the base and produce roots which may descend more than a metre to the ground. Supplejack is restricted to New Zealand, but other species of *Ripogonum* are found in eastern Australia and New Guinea.

The two species of *Parsonsia*, sometimes known as native jasmine, are found in lowland forest and shrubland throughout the country. *Parsonsia heterophylla* is the larger of the two, with stems up to 10 cm in diameter which attain heights up to 20 m above the ground. It is commonest near forest margins, but can reach the crowns of taller trees deeper in the forest by spreading from lower to higher levels in the forest canopy.

There is a very marked difference in size and shape between juvenile and adult leaves in this species (Fig. 34). The seedlings establish them-

*Figure 33 (opposite)* An entanglement of supplejack (*Ripogonum scandens*). Photo: M. D. King.

*Figure 34* The vine *Parsonsia heterophylla*. Juvenile foliage: right. Adult foliage and leaves: left. Photo: J. E. Casey.

selves in sometimes quite shady places on the forest floor and the first few leaves produced are small and almost circular. These are followed by leaves tending towards the second type of juvenile leaf, which is long and narrow with smooth or wavy margins. Intermediates may be narrow at the base and round at the tip, or narrow at the base and tip and round in the middle, and, to make matters even more complicated, lateral branches on the seedlings usually repeat the same sequence. The result is a bewildering and apparently random arrangement of leaf forms. Some of the seedlings are completely green, others are largely brown and some leaves of the latter are unusually attractive with a mosaic of green, dark brown and pale brown patches.[39] The adult leaves, which are formed when the stems reach full light, are much larger and broader and generally uniform in shape, although there are often modest differences in shape between different vines.

The seedlings, which rotate in an anticlockwise direction at their tips as they grow upwards, reach about 45 cm in height without support and somewhat higher if two or more seedlings twine about each other. If no support is encountered then the stems bend down to the ground and grow along it until they find something to climb. The supporting stems are usually slender, although *Parsonsia heterophylla* has been observed climbing tree trunks of up to 25 cm in diameter.

*Parsonsia capsularis* has small flowers, different in form from those of *P. heterophylla,* but the adult leaves of some varieties of the two species are very similar. *P. capsularis* also has long, narrow, reddish-brown juvenile leaves, and in some cases these are retained at the adult flowering stage. This is a smaller plant than its more common relative and grows in shrub communities and, less often, at forest margins. The small, fragrant flowers of these two species are borne in clusters and are white, yellow or, in the case of *P. capsularis* only, red. The fruits are pod-like, hang downwards and split open to release numerous seeds, each with a dense tuft of hairs for wind dispersal.

The distinctive juvenile forms of our parsonsias are not peculiar to New Zealand. The phenomenon is also found in species in eastern Australia and New Caledonia. The genus ranges from tropical Asia to the Pacific.

Two species of *Muehlenbeckia* are twining lianes common throughout New Zealand in lowland to montane forests. *M. australis* is the larger species and its seedlings are often abundant on the forest floor in both shady and well-lit places, but mature plants are most commonly found at forest margins or in regenerating forest. The young stems bend to the ground if they don't find support, branch, and then spread on the forest floor. Unlike most other twiners the erect stems rotate in either direction and in some cases change direction when they begin to climb. The supports are always slender and often become very deformed as they expand within the coils of the vine. Sometimes they die, and this may be caused by the vine, but at other times it is the latter that dies, leaving as evidence on the supporting stem a pronounced helical groove (Fig. 35).

*Figure 35* Stem of a young matai (*Prumnopitys taxifolia*) distorted by a twining vine (possibly *Parsonsia heterophylla*) which subsequently died. Photo: J. W. Dawson.

By climbing up smaller trees *Muehlenbeckia australis* may extend into the crowns of tall trees 30 m or more above the ground. A distinctive feature of *M. australis* is its formation of firm cane-like 'searcher shoots' during the autumn from any part of the stem system. Where these arise on stems coiled on the forest floor they grow erect for several metres, beginning to rotate only after the first metre. Where they develop on stems in the tree crowns they extend more or less horizontally, often from one tree crown to another, and in this way the vines become extremely widespread through the forest canopy. In fact, there sometimes seems to be more of the draping foliage of the *Muehlenbeckia*, particularly in second growth forest, than of the trees themselves. The adult leaves are several centimetres long; broad, thin and pale green, sometimes with a drawn-out tip. Juvenile leaves are much smaller; round, oval or sometimes, fiddle-shaped.

*Muehlenbeckia complexa* is similar in its growth habit to *M. australis,* but it is smaller in all respects and grows on shrubs or small trees at forest margins or in shrub associations. The leaves are only about 1 cm long, more or less circular and a little thicker than those of *M. australis*. The New Zealand species of the genus are largely endemic although *M. australis* occurs on Norfolk Island. The genus also occurs in southern South America and Australia.

*Tecomanthe speciosa* is undoubtedly New Zealand's rarest vine in nature, only one plant having been discovered, on the Three Kings Islands. It has robust twining stems which, in cultivation, extend high into supporting trees. The leaves are pinnately compound with quite large leathery leaflets and the tubular flowers, although large, are of an inconspicuous cream colour. Most of the other species of *Tecomanthe* are found in New Guinea.

Mangemange (*Lygodium articulatum*) belongs to a largely tropical genus of ferns unique because of their ability to climb by twining; although in this case it is not the stems which twine but the axes of compound leaves, which have indefinite growth and sometimes extend from the ground to the tops of high trees. The New Zealand species is restricted to the north of the North Island, but there it is common, particularly in kauri forests. The true stems spread over the forest floor. The axes of the leaves arising from these are slender and wiry and often twine about each other as well as their supports to form springy masses on the ground, in well-lit places or on the forest roof. Compound leaflets arise from the twining axes and, in well-lit situations, these may be fertile with narrow segments, each with two close-set rows of sporangia.

**Twining Leaf Petiole Climbers**

It might be wondered why the fern *Lygodium,* with its twining leaves, is not included here. Leaf climbers are defined as plants whose stems are supported as they grow upwards by the sensitive petioles (stalks) of their otherwise unmodified leaves, which wind round any slender supports with which they make contact. In *Lygodium* the true stems remain on the ground, but the primary axis of the leaf is like a twining stem in its indefinite growth and indefinite ability to twine. Indeed it takes a botanist to appreciate that with *Lygodium* we are dealing with twining leaves rather than twining stems, so it seems more realistic to treat it as a twiner rather than a leaf climber.

New Zealand representatives in this category are all species of *Clematis,* a genus which is widespread in temperate regions and also found in the montane tropics. The best known and largest New Zealand species is *Clematis paniculata,* which is found in lowland forests throughout, particularly marginally, and is greatly appreciated in the spring when its sprays of large, pure-white flowers stand out against the dark foliage of the forest. The adult leaves are divided into three leaflets, which are broad, dark-green, and smooth-margined. Leaves on young plants on the forest floor are very different. The first leaves are long, narrow, membranous in texture and undivided. These are succeeded by compound leaves with three narrow leaflets. In subsequent leaves, the leaflets become deeply lobed, broader, and, where light is adequate, gradually trend to the adult form. Similar juvenile leaves have been reported for *Clematis* species in eastern Australia. Seedlings rotate in an anticlockwise direction and are able to twine around slender supports. Similar twining ability, at least when young, has been recorded for some other leaf climbers outside New Zealand; nevertheless the primary mechanism for climbing is the clasping leaf petiole.

When a leaf is formed at a stem apex it is first erect, then gradually bends downward until it projects at right angles to the stem. The petioles of the leaf as a whole, and of the leaflets, are well developed at this stage and if any of them touch a suitable support they are stimulated, by a process not yet understood, to wind round it. The portion of the petiole in contact with the support enlarges and becomes strengthened. Clearly the leaves cannot attach to large supports, so where a large stem of *Clematis paniculata* up to 10 cm in diameter ascends to a tree crown 10 m or more above the ground, it must have attained that position via smaller trees and shrubs.

The four or five other climbing New Zealand species of *Clematis* are smaller than *C. paniculata*. They are incompletely known and in some cases not yet clearly defined. Several grow at forest margins, including *C. forsteri* and *C. foetida,* and some also grow in shrubland.

### Tendril Climbers

Tendrils are similar to the petioles of leaf climbers in that they are sensitive to touch and respond by twining round a support. They differ in that they are derived from plant organs — branches, inflorescences, leaves or leaflets — which have completely lost their original function and are used solely for climbing. Further, once a tendril has attached to a support, it coils into two opposed helices in its free part, which increases its elasticity and also draws the stem closer to the support.

In New Zealand we have only one forest liane which climbs by tendrils. This is the native passion vine, *Passiflora tetrandra,* which ranges through the North Island and down to Banks Peninsula on the east of the South Island. The leaves are dark green and shiny and drawn out to a point at the tip. The flowers are much smaller and less colourful than those of the cultivated species and less elaborate in their form, but the fruit compensates for this by being bright orange and 2–3 cm in diameter; it is greatly sought after by birds.

The tendrils arise in leaf axils and are considered to be modified inflorescences. They are at first erect then bend downwards; if they encounter a slender support they wind round it. The part in contact gradually becomes thickened, until it is about twice the diameter of the free part of the tendril. The native passion vine is most common in the lower marginal parts of forests, but it spreads so effectively over the forest roof that it frequently reaches the tops of taller trees. The woody stems can be up to 12 cm in diameter and in their lower parts often form tortuous coils on the forest floor.

### Hook Climbers

The New Zealand hook climbers are all species of *Rubus,* a genus which includes the familiar blackberry and raspberry and is widespread in temperate regions and the montane tropics. In north temperate regions, the species of *Rubus* do not climb and are shrubs or scramblers in open habitats; but most of the New Zealand species and a number of Australian and tropical species are low to high climbing forest lianes. The adult leaves of most species are palmately compound with three or more leaflets. Backwardly curving hooks or prickles stud the underside of the petioles and leaflet midribs and sometimes the stem as well — a feature which effectively prevents the stems slipping back from any position attained. In colonial times, this tenacity earned for the New Zealand plants the name of 'bush lawyer', a perhaps unwarranted slur on the legal profession of the day. In fact the 'lawyers' are the only plants in New Zealand forests which are prickly. This is in contrast to the rain forests of Queensland and south-east Asia where many spiny climbers are unpleasantly in evidence. The original lack of browsing mammals in New Zealand is the probable explanation for this, and the same would apply for the lack of spiny plants in New Caledonian forests.

The bush lawyers are sometimes included in the 'scrambler' category of vines. Scramblers are small, unspecialised climbers whose weak, drawn out stems grow up between the branches of shrubs and trail over them. The lawyers begin their ascent in a similar way, but their hooks enable them to reach great heights, equal to those attained by more specialised vines. For this reason I think they warrant a special category. The liane species of New Zealand *Rubus* occur throughout the country, including Stewart Island, in lowland to montane forest.

The first leaves of the young plants are simple; their stems are quite stout and so are able to stand erect without support for 60 cm or more. If nothing is available to climb, the young plant bends to the ground, branches and spreads widely over the forest floor until some of the branches find supports and make their way into the forest canopy via shrubs and smaller trees. The woody stems, which can be looped on the forest floor as well as extending to the forest roof, frequently produce 'searcher shoots' (Fig. 36). The searcher shoots which are near to the ground can stand without support for one metre or more, and are thus very effective in expanding sites of *Rubus* foliage in the canopy and in establishing new sites.

The commonest species is *Rubus cissoides* which has long, narrow and sharply toothed leaflets. The adult stems may be up to 17 cm in diameter and the foliage can reach to 15 m or more above the ground.

*Rubus schmidelioides* has stems up to 10 cm in diameter, and generally smaller, similarly shaped leaflets, but these leaflets are bluntly toothed and have a dense covering of whitish hairs beneath.

*Rubus australis* is most common in swamp forest and is sometimes referred to as 'swamp lawyer'. Its leaflets are short, fairly broad and sometimes almost circular. In this species there is a distinct juvenile form, which spreads and roots widely over the forest floor, bearing leaves with small, membranous, more or less round leaflets with reddish coloured veins. At the adult stage this species can reach for 10 m or more into tree crowns, with stems several centimetres in diameter.

*Rubus squarrosus* is perhaps the most remarkable of the genus, as, in open situations and on shrubs, the leaflet blades remain undeveloped and the leaves consist of rather elongated petioles and the almost threadlike midribs of the leaflets, all beset with yellow prickles (Fig. 37). Such leaves are very effective in clinging to any support. When the stems reach into tree crowns there is a trend towards normal leaves with well formed narrow leaflets. This species is similar in eventual height and stem size to *R. australis*.

*Figure 36* Bush lawyer (*Rubus cissoides*). Horizontal stem near the ground with a vertical stem derived from a searcher shoot.
Photo J. W. Dawson.

*Figure 37 Rubus squarrosus.* Stems, leaf petioles and mostly bladeless leaflets beset with prickles. A few leaflet blades have developed. Photo: J. E. Casey.

## Low Climbers of Forest Margins and Shrublands

Some low climbers of more open habitats — *Parsonsia capsularis, Muehlenbeckia complexa, Clematis* (several species) — have already been considered in association with their higher climbing relatives.

The remaining species are mostly unspecialised 'scramblers', as defined in the last section in the discussion of hook climbers. They grow through shrubs and small trees at forest margins or in shrub communities in drier eastern localities. The species in this group are: *Fuchsia perscandens*; *Brachyglottis (Senecio) sciadophilus* and *Helichrysum dimorphum* of the Compositae or daisy family; *Scandia geniculata* of the Umbelliferae or carrot family (one of the few climbing members of this family); *Carmichaelia kirkii,* one of the leafless brooms; and an undescribed species of *Coprosma*.[40]

One of our species of *Lycopodium,* the attractive trailing *L. volubile,* can also be a scrambling climber over shrubs. This species ranges from New Zealand to the tropical Pacific and south east Asia and, in the latter region, is recorded as sometimes extending into tree crowns.

### Vines Growing as Shrubs in Open Situations

A number of vines can grow as shrubs in open situations where there is no support. This is particularly the case with the small, unspecialised, scrambling species, as well as the small twiners *Muehlenbeckia complexa* and *Parsonsia capsularis,* both of which are able to make an easy transition from well lit forest margin and shrubby habitats to completely open sites. As shrubs these species often have a ball-like form with a profusion of slender, entangled stems.

Some of the taller, more specialised forest vines may also grow as shrubs. *Rubus squarrosus* has been already mentioned. Several of the climbing ratas (*Metrosideros*) can also be encountered as shrubs. *M. perforata* in particular has a dense, billowy form, which gives no hint of any ability to climb. Such dual roles are not peculiar to New Zealand vines, but have also been observed in a number of tropical lianes.[21]

## Epiphytes

Sharing a need for much brighter light than is available on the floor of a closed forest, the majority of species of epiphytes, vines and parasites grow high in tree crowns. In this situation epiphytes alone face special problems. Vines are able to obtain soil water and mineral nutrients via their stems; parasites can tap the supplies drawn up by their host trees; but epiphytes either have no connection with the ground throughout their lives or send roots down to it only after a period of some years. Soil does not form easily on trunks and branches and as they are sunnier, windier and better drained than the ground, they suffer both more frequent and more severe droughts. In response to these stressful conditions, many epiphytes have evolved modifications enabling them to store water and to reduce its loss by evaporation as well as to build up a layer of water-retentive soil. Water can be stored internally in special cells, whose presence confers fleshiness on the organs concerned, or externally in cavities formed by appropriately shaped and arranged leaves. Some epiphytes can get by with a minimum of mineral nutrients and need little or no soil; others build up considerable quantities of dark humus, largely from the decay of their own old leaves and roots with a varying contribution of bark flakes and leaves from surrounding trees. Many other epiphytes which are unable to form soil themselves take advantage of those that can.

Now we have defined the epiphyte category, how rigorously do we interpret the definition in deciding whether or not a particular species should be included? Certainly not so rigorously that we exclude those species which, although normally epiphytic, are sometimes to be found on sunny, rock outcrops which provide conditions similar to those of tree tops. In fact, it might well be that there are no epiphytes, even those of tropical forests, unable to grow on the ground in suitable circumstances.

Going to the other extreme, should we include species which are normally terrestrial, but can occasionally grow on trees? In this case the answer is 'no' as, apart from reservations about stretching the definition so far, the number of species involved would be inconveniently large. In certain circumstances almost any plant is able to grow as an epiphyte. For example, in forests of high rainfall, particularly where frequent mists maintain high atmospheric humidity, seeds germinate just as readily on moist, moss and lichen covered trunks and branches as on the ground. A notable example of chance epiphytism in these circumstances in New Zealand is an occasional silver beech (*Nothofagus menziesii*) growing on a tree of the same species. Even in forests of average rainfall and atmospheric humidity, the branch systems of large, long-lived trees, such as the kauri, are available as habitats for so many centuries that quite unlikely species can sometimes be found as epiphytes on them. The intrepid Harrison-Smith[32] found a 3 m kauri growing on a kauri, as well as several examples of other conifers — rimu, totara, kahikatea and several angiosperm trees. In most cases the occasional epiphytic plants of otherwise terrestrial trees and shrubs are small and do not grow to reproductive maturity.

Lichens, often followed by mosses, are generally the first epiphytes on trees in both temperate and tropical regions. In New Zealand small filmy ferns are frequently associated with the mosses. The thin layer of soil that these small epiphytes form is important for the establishment of most of the vascular epiphytes,[41] which are the concern of this section.

## Shade Epiphytes

These do not require a very high level of light, and as epiphytes they mostly grow low on tree trunks where they escape the shading of the larger plants of the forest floor.

Six small species of fern and two flowering plants occupy this station in New Zealand, although they may also be found on rocks. Of the ferns, three are species of *Grammitis* characterised by tufts of narrow simple leaves arising from short rhizomes. *G. pseudociliata* differs from *G. billardierei* and *G. magellanica* subspecies *nothofageti* in having an abundance of reddish hairs on its leaves. *G. pseudociliata* is concentrated in the North Island; *G. billardierei* and *G. magellanica* extend throughout the country, the last also occurring in south-east Australia.

*Ctenopteris heterophylla* occurs throughout the country as well as in the subantarctic islands and south-east Australia. Its habit is similar to that of the *Grammitis* species, but the leaves are somewhat larger and once-pinnate with toothed leaflets.

*Figure 38 Pyrrosia eleagnifolia,* a fern mat epiphyte. Photo: J. W. Dawson.

*Anarthropteris lanceolata* is found in the North Island and near the northern shores of the South Island, as an epiphyte or on rocks. It is also said to occur in Vanuatu. The leaves are very similar in shape to those of *Grammitis* but longer. The rhizomes are short and produce masses of slender furry roots, some of which can give rise to new tufts of leaves.

These five epiphytic ferns are related to the climbing *Phymatosorus* species and have similar prominent, rounded to oblong, brown to orange sori. The sixth fern which grows as a low epiphyte is a filmy fern, *Hymenophyllum pulcherrimum,* which ranges throughout the country.

The two flowering low epiphytes are both species of *Peperomia,* a large genus of small succulent plants in tropical and subtropical regions. As epiphytes, they grow mostly near the coast in the northern half of the North Island. *P. tetraphylla* ranges from the East Cape district through to the Bay of Plenty. It has leaves in whorls of four as its name indicates. The same species is also found in Australia and Polynesia. *P. urvilleana* bears its leaves singly and grows throughout the North Island and near the northern shores of the South Island, but is less common in the southern part of its range and there grows mostly on rocks. It is found also on Norfolk and Lord Howe Islands.

## Sun Epiphytes

These are most abundant in tree crowns, although some also occur at lower, shadier levels. Sun epiphytes are more numerous and diverse than shade epiphytes and can be grouped into several growth forms.

### Mat Epiphytes

These epiphytes form mats or patches mostly on inclined or horizontal branches, and they comprise three orchids and one fern which mostly range throughout the country. Mat epiphytes may establish directly on bare bark, particularly if it is rough and fissured, but may also avail themselves of moss cushions. The fern is *Pyrrosia eleagnifolia* (Fig. 38), which belongs to a genus of epiphytes centred in tropical Asia. Our species is also found in Australia and the islands of Polynesia. *Pyrrosia* often establishes directly on bare bark and has slender, freely branching rhizomes which form a complete network over sunny branches. The leaves are simple and smooth-margined, varying from almost round to long and narrow. They have a fleshy texture and a dense felt of buff coloured hairs

*Figure 39* Tall kahikatea (*Dacrycarpus dacrydioides*) with many asteliad nests. South of Kaitaia, northern North Island. Photo: B. V. Sneddon.

*Figure 40 (opposite)* Asteliad nests with the pendent epiphytic fern *Asplenium polyodon* below them. Photo: M. D. King.

beneath, which presumably restrict loss of water. *Pyrrosia* also grows on the trunks of trees in the open and on rocks. It can be quite abundant on introduced trees, particularly *Cupressus macrocarpa*.

The epiphytic orchids[42] which form mats or patches belong, or are closely related, to large tropical genera and can be regarded as outliers, reduced in both leaf and flower size. They all have specialised roots which, as well as serving for attachment, also efficiently absorb and store water in a special outer layer of dead cells known as the velamen.

*Drymoanthus adversus,* often attached to quite smooth bark, is unlike the other species in that it has a short stem, which does not grow along the bark surface. The roots arising at the base of the tuft of leaves are particularly conspicuous as they spread out 'like the rays of a spider's web' for a considerable distance, often encountering the roots of other plants of the same species. *Drymoanthus* includes our species, another in east Australia and a third in New Caledonia, but there is some doubt as to whether it should be separated from the Asian and Australian genus *Sarcochilus.* Our two species of *Bulbophyllum,* although small, form quite dense patches with their branching rhizomes. In common with their many tropical relatives, each of their rhizome segments swells at the end into a water-storing 'pseudobulb' with a single leaf arising from the top. The stalk bearing the small flower or flowers arises from below the pseudobulb. *Bulbophyllum pygmaeum* is the smaller species with leaves about a centimetre long but it forms larger patches than those of *B. tuberculatum*. The leaves of *B. tuberculatum* are several times longer, but this species is less frequently seen and is not known further south than the north coast of the South Island.

## Nest Epiphytes

Much more evident to the casual observer are the massive nest epiphytes perched high in tree crowns (Figs. 26, 39, 40). In New Zealand there are three long and narrow-leaved species belonging to two closely related genera of the lily family — *Collospermum hastatum, C. microspermum* and *Astelia solandri.* Two other species of *Collospermum* are found outside New Zealand, one each in Fiji and Samoa; both are epiphytes. In the much larger and more widespread *Astelia* found mostly in the southern hemisphere, only a few species are consistently epiphytic. The others are found in New Zealand in a variety of habitats — coastal cliffs, swamps, forest floors, alpine tussock grassland and a few reduced turf-forming species in alpine bogs.

All three nest epiphytes usually establish among mosses and lichens in branch forks or on inclined or horizontal branches. As their stems, completely hidden by the leaf clusters, are short and more or less erect the plants are fixed in position; although when they branch to form additional leaf clusters the resulting massive clumps of foliage may be metres in diameter. The 'nests' are attached to their supports by extensive root systems and as the old roots and leaves die and decay, considerable depths of dark spongy soil are built up. The likely eventual fate of these large soil and plant masses is to fall to the ground. Heavy rain absorbed by the soil greatly increases the weight of the mass and if the rain is accompanied by wind, complete branches may crash to the ground under the weight of epiphytes.

*Astelia solandri* is more shade tolerant than the collospermums and so is often found below them in the lower crowns and on the upper trunks of trees. The silvery green leaves are in three ranks and are 1–2 m long, but only 2–3 cm wide. Their bases are tightly folded, forming a narrow ridge at the back. *Astelia solandri* is found in lowland forests throughout the North Island and near the northern coast and down the western side of the South Island to about 44°S.

*Collospermum hastatum* accompanies *Astelia solandri* through the North Island and to about 42° 30' in the South Island. *Collospermum microspermum* is restricted to the North Island and replaces *C. hastatum* in montane forests above about 300 m. *C. hastatum* has fan-like arrangements of black-based leaves (Fig. 41) which are somewhat shorter and much broader than those of *Astelia solandri*. The bases of the leaves are strongly rounded and enclose spaces or 'tanks' which become filled with water when it rains. Often sufficient water is contained by the tanks to provide a shower bath for the unwary when a fallen *Collospermum* nest is lifted and tilted. A species of mosquito has been described whose larvae always develop in the water stored by *Collospermum hastatum*[43] but this is a very modest fauna compared with the many insects and even frogs which inhabit the 'tanks' of the tropical American epiphytes of the Bromeliaceae (pineapple family). The water stored by *C. hastatum* is partly absorbed by roots which grow into the tanks, but the suggestion that it is also absorbed through the embedded multicellular bases of overlapping scales has not been confirmed. In some tropical bromeliads all of the stored water is absorbed through the bases of similar scales.

*Collospermum microspermum* is equally specialised but its leaves are as narrow as those of *Astelia solandri* and dark-brown rather than black at the base.

The collospermums are the only known tank epiphytes outside the family Bromeliaceae, but as they form soil very efficiently, they are also nest epiphytes.

**Pendent Epiphytes**

Four mostly New Zealand wide pteridophyte species often grow as epiphytes with their roots or rhizomes embedded in the soil of epiphyte nests. Though they occur elsewhere as well it is in these sites that their growth is most vigorous and their pendulous stems or leaves attain their maximum length. Of the four, *Lycopodium varium* is the most impressive; its slender stems sometimes forming huge masses up to 1.5 m long, below asteliad nests (Fig. 42). The stems branch repeatedly by equal forkings or dichotomies, so that they form a dense but well balanced mass. It is almost constantly in motion as even the lightest breeze can set the tassels swaying. In their upper parts the stems are clothed by small spreading leaves, which grade into small, close-set scales enclosing the sporangia towards the branch tips. *Lycopodium varium* is restricted to New Zealand, but there are related species in tropical forests.

The ferns *Asplenium polyodon* (= *A. falcatum*) and *A. flaccidum* may have leaves of a metre or more in length below the epiphyte nests. *A. polyodon,* with its double-toothed, wedge-shaped leaflets is perhaps the most attractive of the New Zealand species of the genus (Fig. 40). *A. flaccidum* has an unusual stringlike appearance with long and narrow, deeply-toothed leaflets (Fig. 43). Dobbie aptly describes the leaves of this species as appearing to have been 'cut from a piece of pale-green leather'. *Asplenium polyodon* is found from India to Australia and the Pacific and *A. flaccidum* also occurs in Australia. Both aspleniums and *Lycopodium varium* extend into montane cloud forests, but there they depend from mossy trunks and branches.

*Tmesipteris,* a genus restricted to the south west Pacific, is sometimes referred to as a 'living fossil' as it is considered to be one of the most primitive genera of land plants. One of the highlights for botanical visitors to New Zealand is to see a living plant of this genus. *Tmesipteris elongata* subspecies *robusta* has been observed growing from *Collospermum* clumps at a number of localities through the North Island, but not yet

*Figure 41 (opposite above right)* Close view of a *Collospermum hastatum* leaf fan showing the dark leaf bases and roots between the decaying outer leaves. The arrow indicates where water is seeping from the reservoirs between the leaves. Photo: J. W. Dawson.

*Figure 42 (opposite left)* Hanging tassels of *Lycopodium varium* below an asteliad nest. Photo: M. D. King.

*Figure 43 (opposite below right)* Epiphytic plant of the fern *Asplenium flaccidum*. Photo: M. D. King.

in the South Island. Its stems, with their small, simple leaves, are unusually long for a *Tmesipteris* and dichotomise freely. Other species of *Tmesipteris* rarely branch. The smaller *tannensis* hangs from branches.

Three orchid species can also be included as pendent epiphytes. The two earinas belong to a small genus with other species in New Caledonia and Polynesia, but this genus is considered to be closely related to the larger *Epidendrum* of tropical America. Both species have spreading rhizomes and can sometimes extend for several metres along branches. The stems bearing the leaves droop downwards and can be 30 cm or more long. The leaves are formed in two rows, more or less in one plane; those of *E. mucronata* are narrow, thin and quite grasslike while those of *E. autumnalis* are broader and thicker, in keeping with the more robust nature of the plant as a whole. Both species form terminal sprays of small flowers, *E. mucronata* in the spring and *E. autumnalis* in the autumn. The flower clusters of *E. mucronata* hang down and are yellowish orange, those of *E. autumnalis* turn upwards and are waxy white with a strong spicy perfume.

Our sole species of the large tropical genus *Dendrobium* (*D. cunninghamii*) is the largest of New Zealand's epiphytic orchids. Its freely branching stems and narrow leaves form feathery drooping masses. The stems are polished, often bright yellow and very bamboo-like in appearance. The white, reddish-centred flowers are scattered and while modest by tropical standards are, at 2–2.5 cm in diameter, the largest among our epiphytic orchids.

**Small Shrub Epiphytes**

The two species of *Pittosporum* and one each of *Senecio* and *Coprosma* in this category are not usually more than a metre high when growing as epiphytes, but may attain small tree size on the ground. All are endemic to New Zealand.

*Pittosporum cornifolium* is found throughout the North Island and although it is quite a common plant, many people are unaware of its existence, perched as it is inconspicuously in tree crowns. The stems of this plant are spindly and often hang down below the branches. The leaves are thin but firm with prominent veins and the flowers are small and yellowish red. The round, woody seed capsules are a surprise. When they open they reveal a bright red lining and shiny black seeds embedded in sticky, bright yellow fluid.

*Pittosporum kirkii* has a more restricted range, as it is not found further south than the central North Island. It has a more erect growth habit with thicker stems and longer, thicker, almost fleshy leaves with obscure veins. The flowers are bright yellow and the capsules are unusually large (up to 4 cm long), flattened and pod-like. Kirk,[44] after whom the species is named, states that the 'valves contract in a curious manner when the

*Figure 44* The small epiphytic shrub *Brachyglottis kirkii* growing from an asteliad nest. Photo: B. V. Sneddon.

capsule bursts'. The capsule is apparently not so colourful as that of *P. cornifolium*, but is described as having an orange lining.

*Brachyglottis (Senecio) kirkii* is found in lowland forests throughout the North Island but has not been recorded from the South Island. Its growth form has been described as 'candelabra-like' (Fig. 44). The leaves are soft and somewhat fleshy and the flowers, up to 5 cm in diameter, pure white and crowded into dense heads.

The thick and shiny-leaved karamu (*Coprosma lucida*) is best known as a ground plant in shrubby, early forest regrowth on drier sites, but is also reasonably common as an epiphyte in asteliad nests. The species is found throughout the country but presumably is common as an epiphyte only within the range of nest epiphytes.

**Large Shrub Epiphytes**

As well as being larger than those of the preceding category, these also eventually send a root to the ground and so overcome the water supply and soil nutrient problem.

Puka (*Griselinia lucida*)[45] is the most notable in this category. Its large, dark green shining leaves usually contrast so strongly with the foliage of the supporting tree that it stands out even to the casual observer (Fig. 45). Puka is distributed in lowland forests throughout the North and South Islands, but is more common in the north. Its seedlings generally establish in asteliad nests situated at branch forks and its roots ramify through the humic soil. After a few years a strong root begins to grow down the trunk of the supporting tree towards the ground. This root and its branches are closely appressed to the bark of the trunk, and frequently grow into crevices and behind bark flakes. The root tips are white and smooth, but a short distance away from them the root surfaces are often densely clothed with short root hairs. Where the roots are in contact with the trunk, they are anchored by the root hairs and the union is sometimes so complete that when the roots are pulled away they either remove portions of bark or leave strips of their own tissue behind.

Generally, when the root tips reach the ground, one main vertical root enlarges greatly until it attains a diameter of 10 cm or more. This main root usually has a few major branches near the ground and the whole system has a very distinctive appearance resulting from the more or less continuous and pronounced longitudinal grooves and ridges of

*Figure 45 (opposite)* The large shrub epiphyte puka (*Griselinia lucida*) on a kahikatea. The crown of the puka is just below an asteliad nest and its main descending root is to the left of the tree trunk. Te Marua, southern North Island. Photo: M. D. King.

the bark (Figs. 46, 47). In its upper parts the main root gives rise to slender, horizontal, girdling roots (Fig. 48), which often encircle the trunk of the supporting tree many times and so ensure that the puka will not be dislodged even by the strongest gale.

Two other species, *Griselinia littoralis* and *Pseudopanax colensoi,* although mostly terrestrial, can grow as epiphytes in the moist montane or higher latitude forests they favour. When growing as epiphytes they are generally beyond the altitudinal or latitudinal range of asteliad nests and

*Figure 46* Distinctively fluted roots of puka (*Griselinia lucida*) on kohekohes. Waikanae, southern North Island. Photo: M. D. King.

*Figure 47 (opposite)* Roots of puka (*Griselinia lucida*) on tawa (*Beilschmiedia tawa*). The arrow indicates the stem of a white climbing rata (*Metrosideros perforata*). Paraparaumu, southern North Island. Photo: M. D. King.

so establish in the moss and lichen cushions of branch forks. Like the puka, they eventually send one or more roots to the ground.

Broadleaf (*Griselinia littoralis*) is the only other species of its genus in New Zealand. Its leaves are smaller than those of puka, yellowish green and symmetrical or only slightly asymmetrical at the base. In puka, however, the leaf base is very asymmetrical as the two parts of the leaf divided by the midrib are of quite different lengths. Broadleaf has been observed as an epiphyte on a variety of trees. Its descending roots are

*Figure 48 (above)* Puka (*Griselinia lucida*) showing girdling roots. Waikanae, southern North Island. Photo: M. D. King.

*Figure 49 (above)* Mountain five-finger (*Pseudopanax colensoi*) on a kahikatea. National Park, central North Island. Photo: J. W. Dawson.

*Figure 50 (right)* Descending roots and a few girdling roots of the epiphyte northern rata (*Metrosideros robusta*). Te Marua, near Wellington. Photo: M. D. King.

often more massive than those of puka, but they are not grooved. The species ranges throughout New Zealand including Stewart Island. Beyond New Zealand *Griselinia* is found only in Chile, where there are five species, at least some of which are epiphytes.

Mountain five-finger (*Pseudopanax colensoi*) also has a wide range, but is absent north of 36°S and from Stewart Island. I have observed it growing as an epiphyte on kaikawaka or mountain cedar (*Libocedrus bidwillii*) on Mt. Taranaki (Egmont) and on kahikatea (*Dacrycarpus dacrydioides*) on the volcanic plateau near Mt. Ruapehu (Fig. 49).

The descending roots of puka and mountain five-finger seem too slender in relation to their height to stand alone when the supporting trees die, but this may be possible for the more massive roots of the broadleaf.

**Tree or Strangling Epiphytes**

Northern rata (*Metrosideros robusta*)[37] is the most notable and common example here. It is found in lowland forest throughout the North Island and near the north-west coast of the South Island. It is much more frequent as an epiphyte than as a ground plant and it prefers the tall emergent conifers as supporting trees. The earlier stages of its life cycle are very similar to those of the puka. It usually establishes in asteliad nests, although young plants have been observed attached directly to rough bark. A distinctive feature of some small northern rata plants is the development of tuber-like swellings on the roots which, it has been suggested, may serve for water storage.[46] Eventually a root grows down the trunk to the ground giving off horizontal girdling roots at intervals (Fig. 50). Unlike puka this descending root does not remain relatively slender, but gradually enlarges to become a metre or more in diameter. It is often branched near the ground to form a tripod or tetrapod arrangement (Fig. 51). More complicated patterns develop where several branching roots descend from a northern rata crown to form complexes several metres in diameter. In some cases more than one rata may be involved, although this is not easy to determine.

With the development of such a massive root system, when the supporting tree eventually dies the northern rata is able to stand alone on its 'pseudo-trunk' (Figs. 51, 52, 53). If the support was an emergent then the rata now replaces it in that role.

The northern rata and some tropical epiphytic trees of similar habit are often referred to as 'stranglers'. This implies that these epiphytes kill the supporting trees by compressing their trunks within a complete or partial network of roots. Popular writers on New Zealand plants have taken enthusiastically to this idea, describing the northern rata variously as a 'predatory gangster', 'forest bandit' or 'notorious strangler' which 'crushes', 'smothers', 'stifles', or 'squeezes' the supporting tree in an 'iron', 'deadly' or 'fatal' embrace.[47]

Partly as a reaction to these verbal flights, some botanists in recent times have tended to take a contrary view.[48] They point out that the

*Figure 51* Mature northern rata (*Metrosideros robusta*) with a tripod based trunk-like root. The original supporting tree is no longer present. Paraparaumu, southern North Island. Photo: M. D. King.

*Figure 52 (opposite)* Mature northern rata (*Metrosideros robusta*) with a branched trunk-like root system. The original supporting tree is dead, but its trunk persists inside the northern rata roots. The broken top of the trunk is indicated with an arrow. Kaitoke, near Wellington, southern North Island. Photo: M. D. King.

light-demanding northern rata generally establishes in the well lit crowns of mature trees so that, by the time the rata is large enough to stand alone, the supporting tree might well have died of old age. It does seem, however, that the northern rata must have some deleterious effect on the supporting tree through partial overshading, root competition, and perhaps in cases where the supporting trunk becomes enlarged within a well developed rata root cage, through some restriction in the movement of water and nutrients.

Recently a distinctive new tree species of *Metrosideros (M. bartlettii)* has been described.[49] It is restricted to a few forest patches near North Cape and is similar in epiphytic habit to northern rata.

Southern rata (*Metrosideros umbellata*) is rare and localised in the North Island, but quite common in montane and higher latitude lowland forests in the west and south of the South Island and Stewart Island. It is mostly terrestrial, but has been observed growing as a 'strangling' epiphyte in several places. Similar *Metrosideros* epiphytes are known in New Caledonia, Fiji and Hawai'i.

## Epiphytes on Tree Ferns[50]

*Figure 53 (opposite left)* Dead northern rata 'trunk' root, one of whose girdling roots still clasps a portion of the trunk of the original supporting tree. The smaller epiphytes are still living showing that they are not parasites as they are clearly not dependent on their tree supports for nutriment. Kaitoke, near Wellington, southern North Island. Photo: M. D. King.

*Figure 54 (opposite right)* *Dicksonia fibrosa* tree ferns with thick 'skirts' of dead fronds and stout trunks, most of whose bulk is made up by interwoven wiry roots. Inland Taranaki. Photo: J. W. Dawson.

We will consider these separately, as tree ferns provide a substrate rather different from the bark of ordinary trees. To begin with, tree ferns do not branch so only the trunk is available for colonisation. Secondly, the trunks are built up from persistent leaf bases and adventitious roots, which provide a variety of surfaces according to the species, but none is quite like bark. Also the large crowns of leaves cast considerable shade so any epiphytes need to be shade tolerant, at least when young.

The different species of tree fern vary in their suitability for epiphytes. In our largest and most handsome species, the mamaku (*Cyathea medullaris*) which has trunks up to 20 m high and jet-black leaf stalks, the leaf bases decay down to hard leaf scars which collectively form an armour-like surface unsuitable for epiphytes. In the lower parts of the trunks masses of slender roots grow out, adding considerably to the diameter of the trunk, and these too form a hard dry surface.

In the ponga or silver tree fern (*Cyathea dealbata*) the leaf bases decay more gradually and do not form well defined scars. As a consequence, soil forms readily in the interstices and a variety of epiphytes are able to establish. Wheki (*Dicksonia squarrosa*) has a trunk surface similar to that of the ponga. The gully tree fern (*Cyathea cunninghamii*) also has persistent leaf bases, but as these are pressed close to the trunk epiphytes don't establish easily. In *Dicksonia fibrosa* (Fig. 54), *Cyathea smithii* and young plants of *C. medullaris* and *C. cunninghamii,* epiphytes are discouraged in the upper part of the trunk by the persistence of the old leaves as a 'skirt'.[51] Thus epiphytes are most commonly found on the trunks of ponga and wheki.

In moist situations, lichens, mosses, liverworts and smaller and larger filmy ferns may be abundant on tree fern trunks. Climbing ratas and other climbers may also be present as well as a range of seedlings of trees and shrubs which die before reaching maturity. Here we will consider only the consistent and specialised vascular tree fern epiphytes. It is worth noting that the species concerned are different in the main from those occurring on ordinary trees.

### Herbaceous Species

A lycopodium is quite commonly enountered which is smaller than *Lycopodium varium,* and with the leaves associated with the sporangia towards the branch tips often not reduced to scales. Some treat this lycopodium as a distinct species, *L. novae-zelandicum*; others regard it as a form of *L. varium*.

*Asplenium flaccidum,* common on trees, also inhabits tree fern trunks. The filmy ferns *Hymenophyllum lyallii, H. ferrugineum* and *Trichomanes venosum* are almost confined to tree fern trunks and others are frequently present. Several species of *Tmesipteris* remain largely restricted to such sites — *T. elongata* subspecies *elongata* throughout and *T. lanceolata* and *T. sigmatifolia* in the northern North Island.

### 'Stranglers'

One shrub and one tree frequently, and several other species less commonly, play this role on tree ferns.

Five-finger (*Pseudopanax arboreus*) is common as a terrestrial plant in shrubby forest regrowth, but in more mature forest it can be surprisingly frequent as a tree fern epiphyte, mostly on the ponga (Fig. 55), but also on wheki. The seedlings establish at the top of the trunk and, being fairly light-demanding, their leaves soon push between and above the fern fronds. The primary root begins to grow down to the ground, but soon gives off a branch root which grows horizontally around the trunk, sometimes returning to and fusing with the vertical root. It is thus comparable with the girdling roots of the puka and northern rata. The vertical root eventually reaches the ground and sometimes branches to enclose the tree fern trunk in a network of roots near the ground. In the meantime, the crown of the five-finger has continued to branch and grow upward with the tree fern crown following behind it. Raukawa (*Pseudopanax edgerleyi*) and *Coprosma grandifolia* may adopt a similar life style but less frequently.

*Figure 55 (opposite)* Five-finger (*Pseudopanax arboreus*) epiphytic on a tree fern (*Cyathea dealbata*). The root/stem junction of the five-finger is indicated with an arrow. Te Marua. Photo: M. D. King.

Kamahi (*Weinmannia racemosa*), the canopy dominant in many montane and higher latitude forests, also frequently begins its life as a tree fern epiphyte, particularly on wheki (*Dicksonia squarrosa*). In this case the seedlings establish anywhere on the trunks but usually lower down than five-finger. The kamahi sends a root to the ground, which branches several times but it doesn't seem to form girdling roots. Instead it often

sends a branch root vertically upwards within the tree fern trunk. The tree fern continues to grow for a time, but eventually breaks off above the junction with the kamahi, leaving a stump. This stump can often still be discerned among the several spreading trunks of quite large kamahis (Fig. 56). In the north of the North Island *Weinmannia silvicola* and *Ackama rosifolia* may also start their lives as low epiphytes on tree ferns.

## Epiphytes Growing on Rocks

As has already been mentioned, epiphytes sometimes grow on the ground and in some circumstances may be important components of terrestrial communities. For example on Rangitoto Island, a volcanic cone only a few centuries old in Auckland Harbour, the epiphytes northern rata (*Metrosideros robusta*), puka (*Griselinia lucida*), *Brachyglottis (Senecio) kirkii*, *Astelia solandri* and *Collospermum hastatum* are common on dry sunny mounds of scoria. Pohutukawa (*Metrosideros excelsa*) is also present and it hybridises freely with northern rata. Further to the south near Wellington there is a remarkable series of raised beaches where on rocky outcrops a number of normally epiphytic orchids are to be found — *Earina autumnalis, E. mucronata, Dendrobium cunninghamii,* and *Bulbophyllum pygmaeum*. *Drymoanthus adversus* was also recorded early this century, but seems to have disappeared. The ferns *Asplenium flaccidum* and *Pyrrosia eleagnifolia* are also present.

## Epiphytes Growing on Leaves

Epiphyllae, as they are termed, are not so evident on the generally smaller leaves of the New Zealand rain forest as they are on the large leaves of the tropical rain forest. However, in the tropics, the plants concerned are mostly filamentous algae, leafy liverworts and lichens.

One epiphyllous alga in New Zealand is commonly observed on the leaves of mahoe (*Melicytus ramiflorus*). This is a species of *Trentepohlia* and it forms conspicuous reddish patches on old leaves of mahoe in autumn and winter.

Twelve epiphyllous species of liverwort in nine genera have been recorded in New Zealand. They are related to epiphyllae of the tropics and in New Zealand they have been found mostly on fern leaves, but also on leaves of trees and shrubs including horopito (*Pseudowintera*).

Epiphyllous lichens were the subject of a detailed study by Allan[52] at Kitchener Park, Feilding. They were found to be abundant on the leaves of the conifers totara (*Podocarpus totara*), matai (*Prumnopitys taxifolia*) and kahikatea (*Dacrycarpus dacrydioides*), on tawa (*Beilschmiedia tawa*), titoki (*Alectryon excelsus*), rama rama (*Lophomyrtus bullata*), the epiphytic orchid *Earina mucronata* and the liane *Metrosideros colensoi*. The leaves of supplejack (*Ripogonum scandens*), the species of *Coprosma, Pittosporum, Hoheria* and puka (*Griselinia lucida*) were free of lichens. Clearly much still remains to be learnt about leaf epiphytes in New Zealand.

*Figure 56* Multi-trunked kamahi (*Weinmannia racemosa*), with the base of the tree fern on which it is established indicated by an arrow. Photo: J. W. Dawson.

*Figure 57 (right)* Haustoria of a mistletoe (*Peraxilla colensoi*) penetrating into a branch of silver beech (*Nothofagus menziesii*).
Photo: J. W. Dawson.

*Figure 58 (below)* The root parasite *Dactylanthus taylori*. On the right is the knobbly plant body of the parasite with a scaly inflorescence as well as a separate inflorescence. On the left are 'wooden roses' or the distinctive forms of the host roots after the parasites have been removed.
Photo: J. W. Dawson.

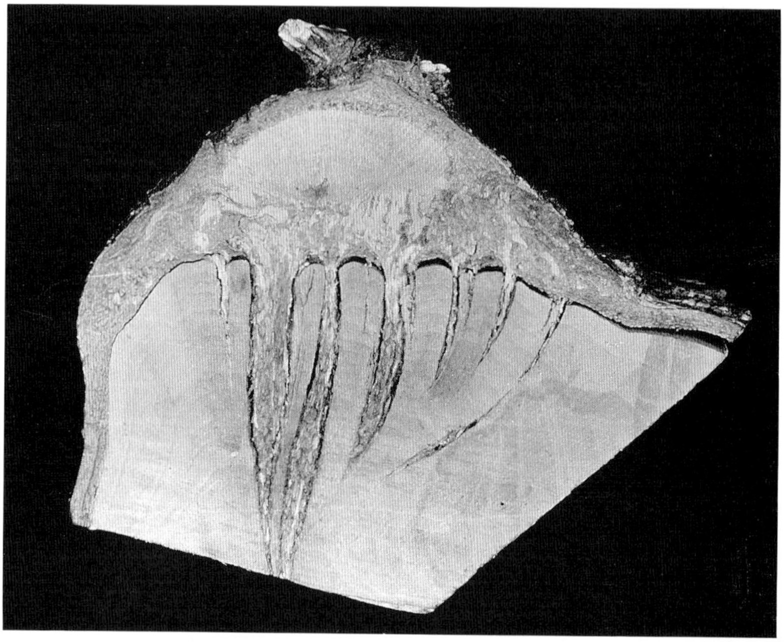

## Parasites[53]

Although all flowering plant parasites agree in having structures known as haustoria (Fig. 57), which penetrate into the living tissues of the host, they differ quite widely in a number of other respects. Some are complete parasites as they lack chlorophyll and so are unable to utilise light energy to manufacture sugars. Others, which have green leaves, make their own organic nutrients and derive from the host mostly water and inorganic nutrients. Some parasites are attached to roots, others to trunks and branches.

### Root Parasites

*Dactylanthus taylorii*[54] (Fig. 58) is a complete parasite attached to the roots of a range of mostly small tree species in lowland to montane forest throughout the North Island. It is not readily observable as only the reddish-brown, scaly inflorescences appear above the ground. The strange appearance of the flower heads which apparently arise directly from the ground, led the Maori to give the name pua reinga (Flower of the Underworld) to this species.

Apparently the embryo root of a *Dactylanthus* seed penetrates the slender root of a suitable host, then gradually expands into a tuber-like structure which eventually surrounds the host root. The terminal portion of the host root then dies away. The 'tuber' continues to enlarge and the end of the host root enlarges with it into a disc-like form. Both can attain a diameter of up to 30 cm. The tuber has a flattened ball-like shape and is covered with hard warty protuberances. Inflorescence buds originate between these, bearing male flowers on some plants, female flowers on others. The flowers have a strong sweet perfume which is attractive to flies. The junction between host and parasite is not flat, but formed into radiating grooves, v-shaped in section. It has been found that if the host/parasite mass is boiled, the parasite can be removed exposing the expanded, fluted ends of the host roots. These 'wooden roses', as they are called, are prized as curios. *Dactylanthus* is restricted to New Zealand but belongs to a largely tropical and subtropical family.

Although *Mida salicifolia*[55] is a root parasite on a wide range of trees including kauri, unlike *Dactylanthus* it does not advertise the fact. It is a small tree with narrow, green willow-like leaves in one variety and rather broader leaves in the other. It is found in lowland forests throughout the North Island, but becomes uncommon in the south of its range. The only other species of the genus is restricted to the Juan Fernandez islands near Chile.

Other root parasites in New Zealand are found in open subalpine and alpine sites. *Exocarpus bidwillii* is leafless with stiff, yellow-green stems branching in a coral-like fashion. Other species of the genus are small trees and shrubs of the tropics (Madagascar, Australia, Malaysia, New Caledonia, Polynesia).

Like *Mida,* the fifteen green-leaved, mostly alpine herbs of the genus *Euphrasia*[55] in New Zealand give no hint of their parasitic behaviour. The genus is widespread in temperate regions of both hemispheres.

## Branch Parasites

These all contain chlorophyll so are only partly dependent on their hosts for organic nutrients. All but one of the New Zealand parasites in this category belong to two largely tropical families — Viscaceae and Loranthaceae, collectively referred to as mistletoes. In the first the flowers are small and inconspicuous, in the second they are much larger and often brilliantly coloured.

Three small species of *Korthalsella* represent the Viscaceae. All have vestigial leaves and strongly jointed stems, which in *K. lindsayi* and *K. clavata* are strongly flattened and in *K. salicornioides,* cylindrical. *K. salicornioides* is found throughout the country while *K. lindsayi* and *K. clavata* are found from the central North Island southwards. Both parasitise a wide range of shrubs and small trees.

The New Zealand species in the family Loranthaceae are all green-leaved, freely branching shrubs of up to one metre in diameter. Currently all these species are referred to a number of small genera,[56] with one exception endemic to New Zealand, although formerly some were included in the tropical genera *Elytranthe* and *Loranthus. Tupeia antarctica* is the only species of a genus restricted to New Zealand. Each plant is attached to a ball-like mass, which is a combination of the parasite haustorium and the host tissues. *Tupeia* is found throughout the country and attacks a wide range of both native and introduced shrubs and small trees, and occasionally other branch parasites.

*Peraxilla colensoi* and *P. tetrapetala* mostly parasitise *Nothofagus* species in both islands. They have bright red flowers which form eye-catching patches of colour against the dark green beech foliage. The orange flowers of *Alepis flavida,* seen mostly on beech trees, are also very attractive. *Trilepidea adamsii,* with reddish cream flowers, is restricted to the northern Coromandel Peninsula and adjacent localities and was last recorded in the 1960s. *Ileostylus micranthus* has small green flowers and yellow berries. It is widespread in New Zealand and Norfolk Island and has as its hosts a range of shrubs, small trees and, sometimes, conifers both native and introduced. The species of all these genera except *Tupeia* send out roots over the bark surface which form secondary haustoria at intervals. *Ileostylus* alone can form new leafy shoots from its roots.

The branch parasites have berries which are eaten by birds and the seeds deposited on tree branches. The seeds are attached to bark by a sticky secretion. Some branch parasites elsewhere have explosive fruits, which shoot the seeds for several metres. This has been observed in the New Zealand species of *Korthalsella*.

The remaining parasite in this category is *Cassytha paniculata*. A twining vine as well as a parasite, it is quite different from the rest. Its seeds

*Figure 59* Tangled thread-like stems of the leafless parasite *Cassytha paniculata* on manuka (*Leptospermum scoparium*). Near North Cape, North Island. Photo: J. W. Dawson.

germinate in the ground and the slender yellow green primary stem with rudimentary leaves rotates in anti-clockwise direction winding tightly around any stems it encounters. At frequent intervals haustoria penetrate the host. The stems of the *Cassytha* branch freely, but remain slender, festooning the shrub hosts with tangled stringlike masses (Fig. 59). This species is restricted to the northern end of the Northland Peninsula where it grows on shrubs and particularly manuka (*Leptospermum scoparium*). It is surprising to find that *Cassytha* belongs to the Lauraceae, a family which otherwise consists of mostly tropical and subtropical trees including the species of *Beilschmiedia* and *Litsea* in New Zealand. Our species of *Cassytha* is also found in Australia and there are other species there and in Melanesia, tropical Asia and South Africa.

## Saprophytes

Vascular saprophytes are completely without chlorophyll and are often small, pale plants growing in leaf litter in very shady places in rain forests. It is thought that they gain their organic nutrients from decaying plant material. Their underground parts are penetrated by fungal threads and recent studies, some in New Zealand, have shown that in some cases the fungal threads are also attached to the roots of nearby trees. It has been suggested that such saprophytes, and probably others, may be secondary parasites drawing nutriment from tree roots via fungal threads.[57,58]

The New Zealand saprophytes are all orchids, except for one belonging to the Burmanniaceae, a family closely related to the Orchidaceae. This is *Thismia rodwayi*,[58] which has been found only in the northern half of the North Island, and there mostly on the volcanic plateau. The pinkish scale-leaved stems, arising from a branching root system, each end in a relatively large delicate flower, which has been likened to a red lantern. Our species is also found in Tasmania and Victoria and there are other species in Australia, tropical Asia and America.

*Corybas cryptanthus* is the only saprophyte among the eight New Zealand species of the genus. It has been collected at scattered localities throughout the country. Only the flower appears above the leaf mould; the stem then elongates to carry the capsule to about 15 cm above the ground. The genus ranges from south-east Asia through Australia to New Zealand.

The 15 species of *Gastrodia* ranging from India and Japan to Australasia are all saprophytes. The branching underground rhizomes are tuberous and filled with starch; those of the New Zealand species were eaten by the Maori. The stems are tall, up to 1 m, and can be attractively if strangely coloured. They often appear polished, with flecks of white and brown which give a resemblance to wood grain. *G. cunninghamii* and *G. minor* are found throughout, *G. cunninghamii* , mostly in beech forest, and *G. minor* mostly under *Leptospermum*. *G. sesamoides* has not been discovered further south than 42°S in the South Island and is found in open forest and shrubland.

*Yoania australis* was only discovered in recent times, but is now known from several localities in *Beilschmiedia tarairi* forest on the Northland peninsula. The stems bearing the small flowers are a pale rose colour and up to 20 cm tall. The genus is entirely saprophytic and is known at several localities in Asia and north Africa.

## Conclusion

In view of its wide range of specialised growth forms and habits, reviewed in this and in the last chapter, it is not difficult to conclude that in these aspects New Zealand conifer broadleaf forest comes closer to tropical rain forest than to any other type of vegetation, despite New Zealand's temperate latitudes. In the light of certain fossil evidence the most likely explanation for this is that, before the Ice Age, forest of the general type now largely confined to tropical latitudes was also widespread in the middle latitudes of both hemispheres. Plant fossils from the vicinity of London dating back to early Tertiary times (80 million years ago) belong to genera, including some palms, now largely restricted to the tropics.[59] Fossil floras with similar relationships have also been discovered in Oregon, U. S. A. The Ice Age, the effects of which would have been more severe in the largely continental northern hemisphere, virtually eliminated such forests from middle northern latitudes, with the exception of southern Japan and parts of China, while limited examples persisted in middle southern latitudes, and in New Zealand best of all.

Remnants of such middle latitude rain forests, with fewer species than those of New Zealand and in particular fewer vines and epiphytes, can be found in parts of New South Wales and Victoria in Australia, along a portion of the south-east coast of South Africa, and in central

Chile. The rain forests of these areas may have been more reduced than those of New Zealand by the development of arid continental climates and by their longer history of human and natural fires as well as by Ice Age coldness. New Zealand's narrow oceanic land mass would have ameliorated the two climatic factors and enabled the survival of our fascinating array of vines, epiphytes, parasites and saprophytes. However, with its fewer species, its relatively small leaves and admixture of conifers the New Zealand conifer broadleaf forest probably comes closest to certain montane tropical rather than lowland tropical rain forests. The conifer broadleaf forest also shares more genera with the former than with the latter.

Perhaps the New Zealand conifer broadleaf forest and related forest types elsewhere in the southern hemisphere evolved in some degree of isolation on Gondwana. The presence of certain specialised growth forms, comparable with but unrelated to those of the old and new world tropics (nest epiphytes — *Collospermum*, *Astelia*; hemiepiphytes — *Griselinia*, *Metrosideros*; lianes — *Metrosideros*, *Muehlenbeckia*) lends support to this suggestion.

## CHAPTER 4

# Conifer Broadleaf Forest: Patterns, Processes and Hypotheses

This chapter will be largely based on the common type of conifer broadleaf forest found throughout the country on adequately drained and reasonably fertile soils. It comprises a mixture of conifers mostly of the family Podocarpaceae; herbaceous and woody flowering plants; and ferns.

The features of other more restricted types will be considered later: well drained coastal forest without conifers; kauri (*Agathis australis*) dominated forest in the warmer northern North Island on less fertile, but well drained soils; mangrove (*Avicennia resinifera*) low forest at similar latitudes in the intertidal zone; and freshwater swamp and bog forests.

## Distribution

At the time of European settlement early last century, conifer broadleaf forest was the most widespread type of vegetation in the North Island (Fig. 1) and also occupied several disjunct areas through the South Island and in Stewart Island. It is believed that before people first set foot in New Zealand about 1100 years ago, conifer broadleaf forest was even more extensive, largely occupying pre-European fire-induced fern and shrubland regions in the North Island and the grasslands and shrublands in the east and south of the South Island. The evidence for this will be reviewed in Chapter 7.

Over the wide latitudinal range of the New Zealand conifer broadleaf forest, from 34 to 47°S (50°S if the southern rata (*Metrosideros umbellata*) forest of the Auckland Islands is included), the total number of species decreases as one goes from north to south, with a tendency towards an aggregation of southern limits at the middle latitude of the North Island, 38°S, and at 42°S in the northern South Island.

## The Commoner Species of the Forest Strata

### Emergents

Conifers are prominent in this usually discontinuous stratum, with rimu (*Dacrydium cupressinum*) the most common in a range of reasonably moist sites on flats, slopes and ridges. The light-demanding and drought-tolerant totara (*Podocarpus totara*) favours stony river terraces and similar level areas, while kahikatea (*Dacrycarpus dacrydioides*) prefers moister places often near streams. Miro (*Prumnopitys ferruginea*), the most shade tolerant of the conifers, and matai (*Prumnopitys taxifolia*) are not quite so tall as the three preceding species and are not always emergent. Miro occupies a similar range of sites to rimu, and matai is most abundant on fertile alluvial or volcanic ash soils. As totara, kahikatea and matai thrive on younger fertile soils they are most prominent during the first centuries of forest development.

Although not tall trees, at higher altitudes Hall's totara (*Podocarpus hallii*) and the attractive conical mountain cedar or kaikawaka (*Libocedrus bidwillii*) may be emergent above a low forest canopy.

Some flowering trees also contribute to this stratum: northern rata (*Metrosideros robusta*) because it commences its life on a rimu or other tall tree; pukatea (*Laurelia novae-zelandiae*) in association with kahikatea in moist places; and rewa rewa (*Knightia excelsa*). The last with its distinctive 'lombardy poplar' form is sometimes abundant on hill slopes, but usually in relatively young forests.

The emergent conifers range throughout the country. Of the three flowering trees rewa rewa and northern rata reach their southern limits in the northern South Island, while pukatea extends to Fiordland on the west.

### Canopy

North of 36°S on the Northland Peninsula, taraire (*Beilschmiedia tarairi*), with its broad mesophyll leaves, dominates the canopy, usually in association with kamahi's northern relative towai (*Weinmannia silvicola*). At higher altitudes in Northland, taraire's relative tawa (*B. tawa*), with its smaller, willow-like leaves, is a minor component of the canopy, but from about 36°S it replaces taraire as the dominant at low altitudes and continues in this role as far as the north-east of the South Island at

*Figure 60* A grove of nikau palms (*Rhopalostylis sapida*) growing near the sea at 42°S on the west coast of the South Island south of Westport. Photo: J. W. Dawson.

42°S. Succeeding tawa altitudinally as the canopy dominant in the North Island from about 39°S is kamahi (*Weinmannia racemosa*); it also replaces tawa in the lowland conifer broadleaf forests of most of the South Island and Stewart Island.

Several other species also contribute to the canopy. Puriri (*Vitex lucens*), which has strong tropical affinities, is limited to the northern half of the North Island; tanekaha or celery pine (*Phyllocladus trichomanoides*) and black maire (*Nestegis cunninghamii*) reach the northern South Island.[60] Hinau (*Elaeocarpus dentatus*) reaches the central South Island, while its higher altitude relative pokaka (*E. hookerianus*) reaches Stewart Island.

On the west and south of the South Island southern rata (*Metrosideros umbellata*) contributes to the canopy of kamahi forests.

### Subcanopy Trees

Two species of decidedly tropical aspect reach the northern South Island, although they are never far from the sea in the southern parts of their ranges. These are kohekohe (*Dysoxylum spectabile*) with its large, pinnately compound leaves and our sole native species of palm the nikau (*Rhopalostylis sapida*)[61] (Fig. 60).

Other species which tend to be more wide ranging are mahoe (*Melicytus ramiflorus*), pigeonwood (*Hedycarya arborea*), toro (*Myrsine salicina*) and two common tree ferns: mamaku or black tree fern (*Cyathea med-*

*Figure 61* The mountain cabbage tree (*Cordyline indivisa*). Photo: J. W. Dawson.

*ullaris*) and ponga or silver tree fern (*Cyathea dealbata*). Some of these species are most abundant in canopy gaps, while other small trees are largely restricted to such sites within the forest; for example, wineberry (*Aristotelia serrata*), putaputaweta (*Carpodetus serratus*), kaikomako (*Pennantia corymbosa*), *Fuchsia excorticata,* lacebark (*Hoheria populnea*), several species of *Pittosporum,* lancewood (*Pseudopanax crassifolius*) and the cabbage tree (*Cordyline australis*).

A few other small trees prefer higher altitudes in northern New Zealand and they include broadleaf (*Griselinia littoralis*) and the mountain cabbage tree (*Cordyline indivisa*). The latter with its unbranched trunk and massive head of broad, silvery-green leaves is certainly the most handsome of our cordylines (Fig. 61). It looks as if it would be at home on a tropical strand, so it seems strange that it should favour moister, cooler, montane forests.

## Shrubs and Small Trees

Kawakawa (*Macropiper excelsum*) is a distinctive undershrub with heart-shaped leaves and jointed stems. Kawakawa is sometimes called 'native pepper tree' because of its hot tasting leaves, and it is in fact related to the true pepper plant of Indonesia. It is also related to the similarly named kava plant of Fiji. Horopito (*Pseudowintera axillaris*) with its dark green shiny leaves has also been termed 'pepper tree' for the same reason, but it is not in fact related to kawakawa. It belongs to the Winteraceae, a family often considered to be the most primitive of the flowering plants.

Other common shrubs are the thin-leaved hangehange (*Geniostoma rupestre*), kanono (*Coprosma grandifolia*) and pate (*Schefflera digitata*) with its large palmately compound leaves. The tree fern, wheki (*Dicksonia squarrosa*), may also be common. It is notable for spreading by horizontal stems or rhizomes to form groves.

In better lit places five-finger (*Pseudopanax arboreus*) may occur. Its leaves are of similar form to those of pate but they have a thicker texture and are more coarsely toothed at the margins. Accompanying species may be the two common larger-leaved coprosmas both known as karamu: *C. robusta* and *C. lucida,* the bubbly-leaved ramarama (*Lophomyrtus bullata*), wharangi (*Melicope ternata*), the 'tree daisies' heketara (*Olearia rani*) and the familiar large-leaved rangiora (*Brachyglottis repanda*).

Among the undershrubs which occur at higher altitudes in the north are mountain horopito (*Pseudowintera colorata*), often with an extremely attractive red or yellow leaf colouration; mountain five-finger (*Pseudopanax colensoi*), *P. simplex*, the tree fern *Cyathea smithii* and *Coprosma foetidissima*.

The coprosma is sometimes known as 'stinkwood' because the crushed leaves smell like rotten cabbage. Indeed the name of the genus is based on this species, 'copros' being latin for dung. Insult is added to injury with the species name, so that *Coprosma foetidissima* could be translated as 'stinking dung plant'. In fact very few of the many species of *Coprosma* have an unpleasant smell.

*Alseuosmia pusilla* is a small shrub which is often overlooked since it frequently grows with mountain horopito and looks very much like it. In the absence of flowers or berries the easiest way to tell them apart is to turn over the leaves — those of horopito are white, those of *Alseuosmia* pale green. It has been suggested that, as the peppery leaves of horopito are unpalatable to deer they may also have been unpalatable to moas.[62] In that case moas, like many bush lovers today, may have passed *Alseuosmia pusilla* by. The genus *Alseuosmia* seems to specialise in such mimicry. I have seen a form of this genus in a forest near Kaitaia with round bullate (bubbly) leaves, and I took it at first to be the familiar ramarama (*Lophomyrtus bullata*).

## Ground Plants

The most abundant plants on the forest floor are ferns, but flowering plants may also be found. Many of the ferns belong to such widespread genera as *Asplenium, Blechnum, Polystichum* and the membranous leaved filmy ferns (*Hymenophyllum* and *Trichomanes*). The two species of *Leptopteris,* although larger than the filmy ferns and not related to them, have similar membranous leaves. The famous crape fern (*L. superba*) favours moist, shady, cool situations and owes its attractively fluffy leaf texture to the ultimate leaf segments, which are set at right angles to the plane of the leaf (Fig. 62).

Among flowering plants, species of *Astelia* form large tussocks; the bush rice grass (*Microlaena avenacea*) covers the ground in places and in season the flowers of species of such orchid genera as *Corybas* and *Pterostylis* make their appearance. In the north, moist shady banks may be completely covered by attractive mosaics of the reddish-purple tinted leaves of parataniwha (*Elatostema rugosum*) (Fig. 63).

*Figure 62 (left)* Crape fern (*Leptopteris superba*). Photo: National Publicity Studios.

*Figure 63 (below)* Parataniwha (*Elatostema rugosum*). In forest south of Kaitaia, far northern North Island.
Photo: J. W. Dawson.

## Life History of the New Zealand Conifer Broadleaf Forest

There are a number of theories on this topic.

**Linear Succession (Climax)** (Fig. 64)

*Figure 64* Diagrams illustrating linear and cyclic forest successions. (Reproduced with permission from *To Save a Forest — Whirinaki.*)

A postulated sequence[63,64,65] which leads to the present forests, with their overstorey of scattered podocarps and canopy of flowering trees, is as follows:

Where the previous forest cover is completely removed by such agents as volcanism or fire, the first coloniser is bracken fern (*Pteridium*

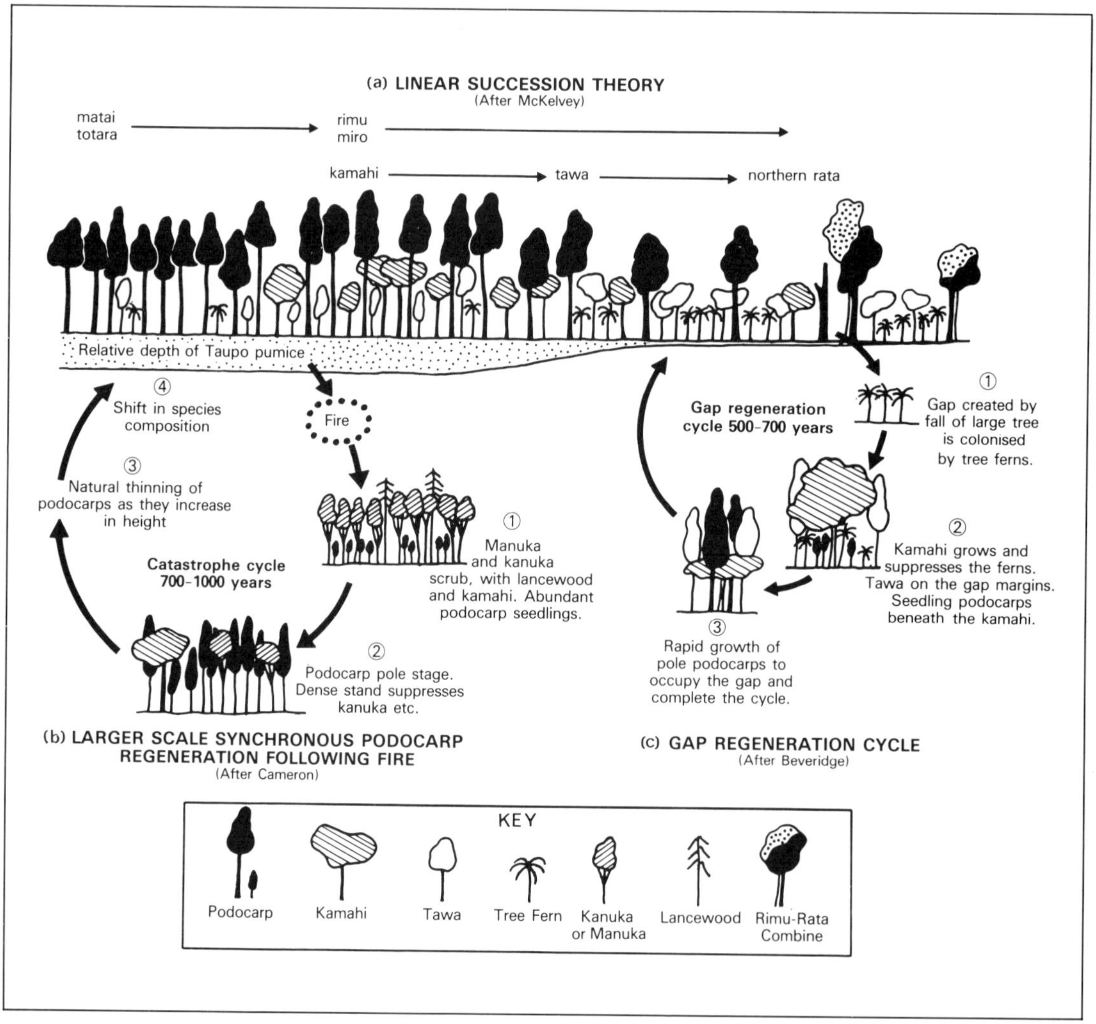

*esculentum*) followed in order by two small-leaved light demanding species — the shrub manuka (*Leptospermum scoparium*) and the small tree kanuka (*Kunzea ericoides*)

Within this association, broader-leaved shrubs and small trees establish, including five-finger (*Pseudopanax arboreus*), lancewood (*P. crassifolius*), *Coprosma lucida, Pittosporum tenuifolium,* juvenile rewa rewa (*Knightia excelsa*), juvenile kamahi (*Weinmannia racemosa*), and in moister places the tree ferns mamaku (*Cyathea medullaris*) and ponga (*C. dealbata*), mahoe (*Melicytus ramiflorus*), wineberry (*Aristotelia serrata*), tree fuchsia (*F. excorticata*) and pate (*Schefflera digitata*).

When these broad-leaved plants form a canopy the bracken, manuka and kanuka gradually die out as the light falls below the level required for their continued recruitment. It is at this stage, about 50 years from the beginning of the sequence, when the forest floor is sheltered but still quite well-lit, that the podocarp conifers become established. After a further 50 years or so the podocarps begin to grow above the broadleaf canopy and by the time they form a higher stratum themselves, the lower layer of flowering trees is now dominated by kamahi (*Weinmannia racemosa*), often initially epiphytic on the ponga tree fern, with scattered rewa rewas (*Knightia excelsa*), hinau (*Elaeocarpus dentatus*) and maire (*Nestegis cunninghamii* and *N. lanceolata*).

Finally, shade-demanding species enter and tawa (*Beilschmiedia tawa*) largely replaces kamahi in the forest canopy. The forest floor is now strongly shaded and the podocarps, light-demanding to varying degrees, no longer establish. The podocarp trees which are already present, often in combination with epiphytic northern ratas (*Metrosideros robusta*), gradually die out and the climax, according to this hypothesis, is a forest with the canopy dominated by tawa and without emergents. The tawa and other shade-demanding species are able to maintain themselves indefinitely until some catastrophic event destroys the forest to initiate a new succession.

**Cycles** (Fig. 64)

Field observations by a number of botanists have led to some questioning of the above sequence of events. Some have suggested that, rather than there being a succession leading to a climax, there may in fact be repetitive cycles. For example, on the volcanic plateau of the central North Island it has been observed in forests with emergent podocarps over a main canopy of kamahi, or kamahi and tawa, that where a podocarp falls the gap is occupied by tree ferns. Kamahi establishes epiphytically on the tree ferns and podocarps then regenerate under the kamahi and eventually overtop them.[56,67] A similar sequence has been observed in south Westland.[68]

Probably because of their importance as timber trees, the lack of regeneration of podocarps in some New Zealand conifer broadleaf forests has received particular attention. According to the successional and cyclic hypotheses just reviewed this would be a normal consequence of forest development.

### Climate Change

Holloway[69] suggested that the general lack of regeneration of podocarps since the time when the present mature to aging emergents established is due to a climate change to colder and drier conditions from about 1300 A. D.

Holloway based his climate change hypothesis on forest patterns in western Southland, in which he saw evidence of a downward movement of vegetational zones, and on historical and other evidence of a worldwide colder interval, the 'Little Ice Age', during the last millenium.

Wardle[70] investigated a number of podocarp-dominated stands on the west and east of the South Island and in Stewart Island. He supported Holloway's hypothesis but concluded that the 'regeneration gap' started later than Holloway suggested, from about 1600–1800 A. D., and that it was most marked in the drier eastern South Island and least marked on the wetter west and in Stewart Island. Since about 1800 the regeneration of podocarps has resumed at some localities.

### Angiosperm Forest Dominance

According to a more dramatic interpretation of the dynamics of the New Zealand conifer broadleaf forest put forward by Robbins,[71] it is not in fact one forest, but two in competition — one comprising the conifers and the other the flowering trees. The conifers belong to a more ancient and less specialised group of plants, which along with ferns and their allies formed forests in New Zealand before flowering plants became dominant throughout the world. Following the establishment of the more specialised flowering trees in New Zealand the less specialised conifers, he suggests, have been on the road to extinction.

Their present poor regeneration is seen as a reflection of this trend. Only where raw new soil conditions, unsuitable for most flowering trees, are brought about by volcanism or glaciation can the conifers still form dense forests and even these give way to flowering trees as more mature soils develop.

The validity of this interpretation can be questioned. Firstly, conifers have coexisted with angiosperms in New Zealand for 100 million years, so it seems unlikely that they will disappear for some time to come. Secondly, if the conifers are doomed to extinction because they are more ancient and less specialised, wouldn't this be even more true for the ferns, an older and less specialised group than the conifers? In fact ferns are abundant in New Zealand forests and give no cause for any belief that they are on the road to extinction. The probable truth of the matter is that when a more specialised plant or animal group becomes dominant throughout the world, many members of the preceding dominant group become extinct; but there is no reason why the survivors could not evolve new forms suited to the changed conditions. This would certainly appear to be true for the ferns as most forest ferns belong to an advanced group, which came into existence at about the same time as the now dominant angiosperms and diversified with them. The tree ferns and filmy ferns are more ancient groups but they give every indication of being permanent components of the conifer broadleaf forests.

As far as conifers are concerned New Zealand is not the only place where apparently inadequate regeneration in mature forests has been noted. It has been observed in forests of Melanesia with species of *Araucaria*[72] and *Agathis*,[73] and similar extinction hypotheses have been proposed. Studies in both areas have been carried out to test the validity of these hypotheses and they have all concluded that the conifers concerned have a permanent role in the forests as a result of recurring natural disturbances. In New Zealand a similar investigation[74] has been made into the role of the mountain cedar (*Libocedrus bidwillii*) in a number of montane rain forests in the South Island and the same conclusion has been drawn.

This seems to suggest then that the failure of emergent conifers (as indeed of many tropical angiosperm emergents[75,76]) to regenerate under a dense canopy is a normal feature of rain forest development, and that special climatic and evolutionary hypotheses are unnecessary.

## More Restricted Forest Types

### Coastal Forest

This type is not greatly different from the general conifer broadleaf forest of better drained sites. It is best developed along the Northland coasts and adjacent islands and is dominated by three species that rarely occur very far from the sea. Pohutukawa (*Metrosideros excelsa*), which often grows alone on coastal cliffs, is restricted to the northern half of the North Island, but karaka (*Corynocarpus laevigatus*) with its large, dark green leathery leaves and ngaio (*Myoporum laetum*) extend to the northern and eastern South Island.

Coastal forest is of lower stature than inland forest and, as a result of the general absence of conifers, it lacks emergents.

In addition to the three trees already mentioned, of which pohutukawa is the largest (and when in flower the most spectacular with its bright red stamens) there may be a number of other trees, shrubs and herbs. Many of these also occur in inland forests, at least in the far north. Notable among these are the trees puriri (*Vitex lucens*), kohekohe (*Dysoxylum spectabile*) and the shrub kawakawa (*Macropiper excelsum*).

Of particular note are a number of large- to very large-leaved species in the northern North Island which range from moderately common to rare. Some of them have strong tropical affinities, and so can be regarded as relicts from warmer times.

On the mainland as well as on islands, and extending to East Cape and beyond, are the small trees parapara or the 'bird catching tree' (*Pisonia brunoniana*) and tawapou (*Planchonella costata*). Parapara has extremely sticky fruits to which small birds can become attached. It is also found in Australia and some Pacific Islands. The tawapou is closely related to species in Norfolk Island and Fiji.

The remaining species are found on the Three Kings Islands[85,86] off the northern tip of New Zealand and are either restricted thereto or also occur on the Hen and Chicken Islands or Poor Knights Islands further south.

(a) Genera with no other New Zealand species:
  Pukanui (*Meryta sinclairii*): a small tree with very large simple leaves which is also found on the Hen and Chicken Islands off the Northland east coast. *Meryta* is centred in New Caledonia with a few species elsewhere in the Pacific.

*Tecomanthe speciosa*: a liane with compound leaves and cream-coloured tubular flowers. Only one plant is known. Other species of the genus are found in Queensland and New Guinea.
*Elingamita johnsonii*: an endemic genus related to *Tapeinosperma* of the tropics. Only one tree is known.
*Davallia tasmanii:* a fern belonging to a largely tropical genus.

(b) Genera with one or two more widespread species in New Zealand:
*Pennantia baylisiana*: one tree only is known.
*Alectryon grandis*: one tree only is known on the Three Kings. It is possibly also found on the Poor Knights Islands.[87]
*Streblus (Paratrophis) smithii.*
*Cordyline kaspar:* also on Poor Knights Islands.[87]
(Some of the rare species may have been more common before goats, which have since been exterminated, were released on the Three Kings Islands.)

It should not be thought that plants are restricted to coastal forest because they require a salty environment. They grow near the coasts because that is where the mildest climates are. If New Zealand extended further to the north, then many of them would occur at inland sites, and indeed the nearest relatives of pohutukawa in the tropical Pacific are not coastal at all, but are found in mountain forests. Conversely, in the far south of New Zealand some species, of inland forests further north, can be found growing close to the sea.

**Kauri Forest**

The kauri (*Agathis australis*) is one of the world's largest trees with an excellent timber that was extensively exploited following European settlement. Young trees are narrowly conical, but mature trees have widely branching crowns and huge cylindrical trunks with little or no reduction in diameter with height.

Although fossils of the kauri which date back to warmer geological times have been found in the far south of the country, at present it reaches its limit at about 38°S, so kauri forests are essentially a feature of the Northland and Coromandel peninsulas. Not all forests in these far northern areas, however, are dominated by kauri.

On more fertile soils, such as those derived from basalt, conifer broadleaf forest of the general type we have already considered prevails, while kauri forest is largely restricted to the less fertile soils derived from consolidated sand dunes, claystone and the sandstone known as greywacke. In the latter case kauris tend to be concentrated on the thinner soils of ridges and spurs with ordinary conifer broadleaf forest in the valleys.

The kauris with their great trunk columns are the overwhelming feature of these forests (Fig. 65). Their crowns form a high canopy at about 35–40 m and plant growth below this is often rather sparse. Most of the tree species of the non-kauri forests are present however, including rimu, taraire, tawa, towai and northern rata, but they tend to be rather small-crowned and spindly. Most of the shrubs and ground plants are found elsewhere, although some are particularly abundant in kauri forest: toru (*Toronia toru*), neinei (*Dracophyllum latifolium*) mairehau (*Phebalium nudum*) and a terrestrial form[77] of the normally epiphytic *Brachyglottis (Senecio) kirkii*. Particularly conspicuous where they occur are the often huge tussocks of 'kauri grass' (*Astelia trinervia*) and the sedge *Gahnia xanthocarpa*.

The fallen leaves, bark flakes and twigs of the kauris form a litter poor in nutrients, which is slowly decayed by fungi rather than the more nutrient-demanding bacteria. In these circumstances the litter accumulates to considerable depths, sometimes up to 3 m near large trees. The litter becomes very acidic and this promotes heavy leaching from the upper layers of the soil and the formation of a thick concrete-like iron pan at a lower level. The impervious nature of this can result in quite swampy conditions.

Kauris not only establish on less fertile soils; they also greatly impoverish them. It is not surprising, then, that there is little regeneration of kauris on the floor of a kauri forest. Cockayne[63] thought that this was due to there being insufficient light on the forest floor, but more recent studies suggest that it results from the infertility of the litter and soil and perhaps also to competition from the roots of the mature trees.[78] Whatever the cause, the lack of regeneration shows that kauris are unable to replace themselves in closed forests, and Cockayne postulated that the climax forest would be one without kauris in which the dominance would probably be assumed by taraire.

However, the study of soils under some kauri forests has revealed a succession of 'fossil soils' at depth which contain kauri gum and other remains. This demonstrates a succession of kauri forests on the same site. As each generation of kauris died out, perhaps as a result of

*Figure 65 (opposite)* Kauri (*Agathis australis*) forest interior. The large trunks are kauris. The narrow-leaved shrubs are neinei (*Dracophyllum latifolium*), and the narrow-leaved ground herbs kauri grass (*Astelia trinervia*). Puketi forest near the Bay of Islands, northern North Island. Photo: J. W. Dawson.

conditions they themselves induced, the various flowering trees and shrubs remaining or following would restore soil fertility sufficiently to permit the establishment of a new kauri generation.

In some cases buried kauri logs are all aligned in the same direction, indicating destruction by a cyclone. Bieleski[78] suggests that the upturned roots of the blown over kauris would break through the iron pan, improve the drainage of the soil and so allow a new sequence to commence.

At present there is extensive kauri regeneration in areas where forests have been logged or burnt in European times. The pioneer phase is dominated by the light-demanding manuka (*Leptospermum scoparium*) followed by kanuka (*Kunzea ericoides*). Numerous kauri seedlings and those of other species develop under the kanuka canopy and as the latter opens out the kauris grow through and overtop it.

Another type of shrubland widespread in the north is found on terrain known as 'gumland'.[79] The kauri forests preceding the gumland vegetation left in the soil considerable quantities of resin, which was gathered intensively following European settlement for the making of polishes and varnishes. The soil is highly infertile and it is thought that repeated fires in both Maori and European times have led to the present degraded vegetation. Manuka (*Leptospermum scoparium*) is quite common, although stunted, and other characteristic shrubs are *Dracophyllum lessonianum* and the strikingly yellow-flowered kumarahou or 'Gumdigger's Soap'[80] (*Pomaderris kumeraho*). Among herbs, sedges and the umbrella fern (*Gleichenia dicarpa*) are prominent. Undoubtedly the gumlands would have to be free of fire for a very long time before kauri forests could return.

Finally, the most puzzling sites containing kauri remains are swamps, as kauris do not grow in such sites except sometimes marginally. Not only is it common to find kauri gum in Northland swamps, but it is often located in several distinct layers indicating a succession of forests, presumably at times when drainage was good, alternating with treeless swamps.

Cheeseman,[81] when considering the swamps of the low lying sandy peninsula leading to North Cape, was the first to suggest an explanation involving cyclical changes in height of the land above the sea. He was thinking in terms of raising and lowering of the land, but, although the effect was the same, we now know that it was a matter of raising and lowering of the sea level. At each glacial period during the recent (or perhaps current) ice age, sea levels dropped by up to 100 m as large amounts of water were locked up in ice. Then, with each interglacial, ice would melt to varying extents and the sea level would rise again. With the lowered sea level, drainage of the swamps would improve and kauri forest would establish; with a raised sea level the forests would disappear as drainage deteriorated sometimes to the extent of inundation by the sea.

If this explanation for the former existence of kauri forests on currently swampy sites is correct it presents us with a curious contradiction: on such sites the kauri, now restricted to the warmest part of New Zealand, flourished during glaciations and disappeared during interglacials!

## Mangrove Forest

Mangrove is the term applied to a number of small trees, mostly tropical, which grow between the tides in shallow seas bordering sheltered coasts. In New Zealand we have only one species, *Avicennia resinifera*. The derivation of the species' name is interesting; it was applied by the botanists on Cook's first voyage, in the mistaken belief that the bits of floating kauri gum they observed came from nearby mangroves.

Like the kauri, the mangrove is restricted to the warmest part of the country reaching the Bay of Plenty in the east and Kawhia in the west. Towards its southern limits it is no more than a shrub, but further north the mangrove can be a tree up to 10 m high forming low forests with a close canopy. For the most part these forests contain no other species of flowering plants.

## Fresh Water Swamp Forest

In the lowlands throughout New Zealand, swamp forest often occurs in localised areas depressed below the general level of the landscape. Such depressions can be found in the irregular terrain of volcanic and formerly glaciated areas; in areas where land blocks have sunk along faults; where meandering rivers have changed their courses; and between consolidated sand dunes.

If there is some depth of water initially, then water plants begin the succession. Some of these are completely submerged and grow on the bottom of the pond or lake; others are free floating at the surface. Where the latter grow near the shore they gradually reduce the water depth with the accumulation of their own remains and the stream-borne silt they trap. In the shallower water which results, some aquatic plants with submerged roots but foliage raised well above the water enter the progression. Notable among these is the raupo (*Typha orientalis*). The raupo in turn raises the ground level to coincide (more or less) with the water level and then gradually gives way to the so-called New Zealand flax (*Phormium tenax*) intermixed with toetoe (*Cortaderia toetoe*) and *Carex secta*. This stage is an impressive sight when the *Phormium* is bearing its spiky reddish flower heads and the toetoe its creamy yellow-

*Figure 66* Swamp with cabbage trees (*Cordyline australis*) and New Zealand 'flax' (*Phormium tenax*). Near Kaeo, northern North Island. Photo: J. W. Dawson.

*Figure 67 (opposite)* Exterior view of kahikatea (*Dacrycarpus dacrydioides*) swamp forest. Harihari, west coast, South Island. Photo: J. H. Johns.

ish plumes. In time shrubs tolerant of poorly drained soil become established — manuka (*Leptospermum scoparium*), koromiko (*Hebe stricta*), karamu (*Coprosma robusta*) and a variety of divaricating shrubs including several coprosmas. The first trees to appear are often cabbage trees (*Cordyline australis*) (Fig. 66), so characteristic of many New Zealand landscapes. Finally come the trees of the closed forest, the most prominent of these being kahikatea (*Dacrycarpus dacrydioides*). In kahikatea swamp forest (Fig. 67), as it is often called, the trunks of the dominant trees may be close-set and exceedingly tall with relatively small greyish-

green crowns. In the North Island and northern parts of the South Island, pukatea (*Laurelia novaezelandiae*) is a common associate of kahikatea and in this type of forest often has well-developed plank buttresses which extend into roots raised above the forest floor with shield-like pneumatophores here and there. The swamp maire (*Szygium maire*) has a similar distribution, but it is a smaller tree.

There is an often rather open undergrowth of small-leaved and large-leaved shrubs, many of which persist from earlier stages in the succession, and sedges are prominent among the forest floor plants. Herbaceous and woody epiphytes are often common and among the vines kie-kie (*Freycinetia baueriana* var. *banksii*) is particularly abundant, covering the forest floor in places and completely obscuring the trunks of trees up to a considerable height (Fig. 32).

Over time, when the soil has built up sufficiently above the water table, the swamp forest will give way to ordinary conifer broadleaf forest.

### Bog Forest

Bogs, although similar to swamps in that they are waterlogged, are much less fertile and much more acid, and the slow breakdown of plant remains which results from these conditions leads to the build up of peat, often with peat moss (*Sphagnum*) playing a prominent role. The greater fertility of swamps is due to the fact that most of their water content comes either from streams or from ponds or lakes fed by streams. The latter contain minerals dissolved from the rocks and soils through which they pass and in times of flood they may also carry quantities of fertile silt. Bogs on the other hand are found mostly in high rainfall areas and most of their water comes as rain, which is devoid of nutrients. Their other characteristics follow from this.

Raised bogs have convex surfaces and consequent outward drainage. They are best developed in the Waikato and parts of Southland and they may become established on former swamps.

In the North Island forested bogs are largely restricted to middle altitudes on the central volcanic plateau near to Mt. Ruapehu. In the South Island they are common on the lowlands of the wet western side and also in Southland and Stewart Island.

Bog forest is low in stature, up to 15 m, and the usual dominant tree is silver pine (*Lagarostrobos colensoi*)[82] intermixed with small trees of the

*Figure 68 (opposite)* Aerial view of the transition from pakihi bog through silver pine (*Lagarostrobos colensoi*)-dominated forest to rimu (*Dacrydium cupressinum*)-dominated forest on better drained terrain. South Westland.
Photo: J. H. Johns.

mountain celery pine (*Phyllocladus aspleniifolius* var. *alpinus*). *Neomyrtus pedunculata* may be prominent in the subcanopy and large tussocks of *Astelia fragrans* may be abundant on the forest floor. The progression from bog to bog forest has been studied in South Westland where island-like patches of bog, known locally as pakihi, are scattered in forest on the lowland gravel plain. The pakihi vegetation is dominated by the rush-like *Empodisma minus* with the sedge *Baumea teretifolia,* the umbrella fern *Gleichenia dicarpa* and various other small herbs. Stunted shrubs of manuka and *Dracophyllum longifolium* are scattered throughout. In the gradation from bog to forest there is firstly a zone of manuka of small tree size under which silver pine in particular establishes. In the next zone, silver pine has formed a canopy above the manuka. Silver pine is itself overtopped by rimu (*Dacrydium cupressinum*) in the next zone (Fig. 68).

As with kahikatea swamp forest, it seems probable that as the soil level builds up and soil moisture declines sufficiently, the progression ultimately leads to the general conifer broadleaf forest of the region. Any trend towards increased regional rainfall would, of course, slow or even reverse such a progression.

Mark and Smith[83] concluded that the pakihi in their south Westland study area had never previously supported forest, but Rigg,[84] who studied pakihis in north Westland, suggested, on the evidence of buried logs, that forests (probably dominated by rimu) did formerly occupy such terrain and that there was possibly a succession of such forests. The soil of these pakihis is highly leached and infertile with a well-developed iron pan, which also suggests a previous forest which provided an acid litter. Rigg proposes, as one possibility, that the podocarp forests would have died out as a result of the reduced soil fertility and, presumably, the impeded drainage they themselves induced. Between successive forests a regeneration sequence similar to that described for South Westland would have taken place. This hypothesised forest-bog-forest cycle is similar to the forest/swamp/forest cycle proposed for some kauri localities.

As with the gumlands of Northland, many of the pakihis of Westland have been repeatedly burnt in both Maori and European times.

CHAPTER 5

# Beech Forests

The second major forest type[88] in New Zealand provides a strong contrast to the conifer broadleaf forest. It is dominated in the canopy or forest 'roof' by one or more species of southern beech (*Nothofagus*),[89] a genus restricted to the southern hemisphere, but nevertheless related to the oaks and true beeches (*Fagus*) of the northern hemisphere. *Fagus* and *Nothofagus* are largely distinguished by flower and fruit characteristics, but in addition the New Zealand species of *Nothofagus,* and a number elsewhere, differ from *Fagus* in being evergreen and having much smaller leaves. However, despite the clear differences between the two genera, it has become customary in New Zealand to refer to our species as 'beech' without qualification.

## Species and Varieties of New Zealand Beech

Five species of *Nothofagus* were formerly recognised in New Zealand, but more recently two of these have been combined as varieties of one species.[90]

*Nothofagus menziesii*: silver beech (the bark is white and silvery particularly in young trees).

*Nothofagus fusca*: red beech (the timber is reddish in colour).

*Nothofagus truncata*: hard beech (the wood has a high silica content making it hard).

*Nothofagus solandri* var. *solandri*: black beech (the bark is very dark partly due to the growth of a sooty mould).

*Nothofagus solandri* var. *cliffortioides*: mountain beech (grows mainly at higher altitudes than the other beeches).

Silver beech has thick, almost circular leaves, 0.5 1.5 cm long and wide, bearing a few domatia on the underside and small, rounded marginal teeth often in pairs (Fig. 69). The leaves of red and hard beech are similar to each other, 2–4 cm long by 1.5–2.5 cm wide and with saw-like marginal teeth. Hard beech has rather thicker leaves with 8–12 teeth per side compared with 6–8 for red beech; red beech alone has domatia. Black and mountain beech have small smooth-margined leaves, hairy below, 1–1.5 cm long by 0.5–1 cm wide. It is difficult to distinguish the leaves of the last two forms. Mountain beech at higher altitudes often has leaves that taper from the base, while the leaves of black beech at low altitudes are oblong, but there are intergrading forms in many places.

Silver, hard and red beech may become large trees 30 m or more in height with trunks 2 m or more in diameter. The trunk bases of the

*Figure 69* Silver beech foliage and young fruits with glandular hairs. Photo: M. D. King.

first two have low, rounded buttresses while those of the latter are high and narrow. The larger trees of mountain and black beech are generally below 25 m in height with unbuttressed trunks of about 1 m in diameter.

The bark of all the beeches, except when young, is rough and fissured. The tree crowns, especially in more open situations, are wide and deep and particularly in mountain and black beech, the foliage is often arranged in distinctive horizontal layers. The male and female flowers at the tips of twigs are wind-pollinated and inconspicuous, although in good flowering years coloured anthers in the male flowers may impart a reddish glow to the crowns of black and mountain beech particularly. More brilliant splashes of colour, contrasting with the dark green of the foliage, may be provided by certain mistletoes which commonly parasitise the beeches. *Peraxilla colensoi,* usually found on silver beech, and *P. tetrapetala* have bright red flowers. *Alepis flavida* with orange-yellow to yellow flowers used to be common, but has now been depleted by opossums. Heavy flowering of all the beech species in New Zealand takes place every few years with little or no flowering in the other years. It is suggested that 'flowering years' follow hot summers which encourage the initiation of flower buds.[91]

The beech seeds are 5–8 mm long with narrow wings. It is considered that they are not capable of being dispersed for more than a few kilometres and then only with strong winds.[92]

Silver beech stands apart from the other species in several respects. Some of the more striking differences are:

(a) each leaf of a silver beech generally lasts several years before falling, those of the other species last for only one year, generally falling a few weeks after the new season's leaves have unfolded, or even, in the cases of occasional trees of red and hard beech, a few weeks before;

(b) natural hybrids have often been observed involving all species except silver beech;

(c) silver beech is the only species attacked by a fungus, *Cyttaria,* whose orangey, golf ball like fruiting bodies may be abundant on the branches of some trees;

(d) study of the pollen of the New Zealand beeches has shown that silver beech has a different form of grain from that shared by the other species.

For the last reason our species have been assigned to two informal groups within the genus, both of which are also represented in Australia and South America — the 'menziesii group' and the 'fusca group'. A third 'brassii group', with a third pollen type, is restricted at present to New Caledonia (Fig. 70) and New Guinea.[93]

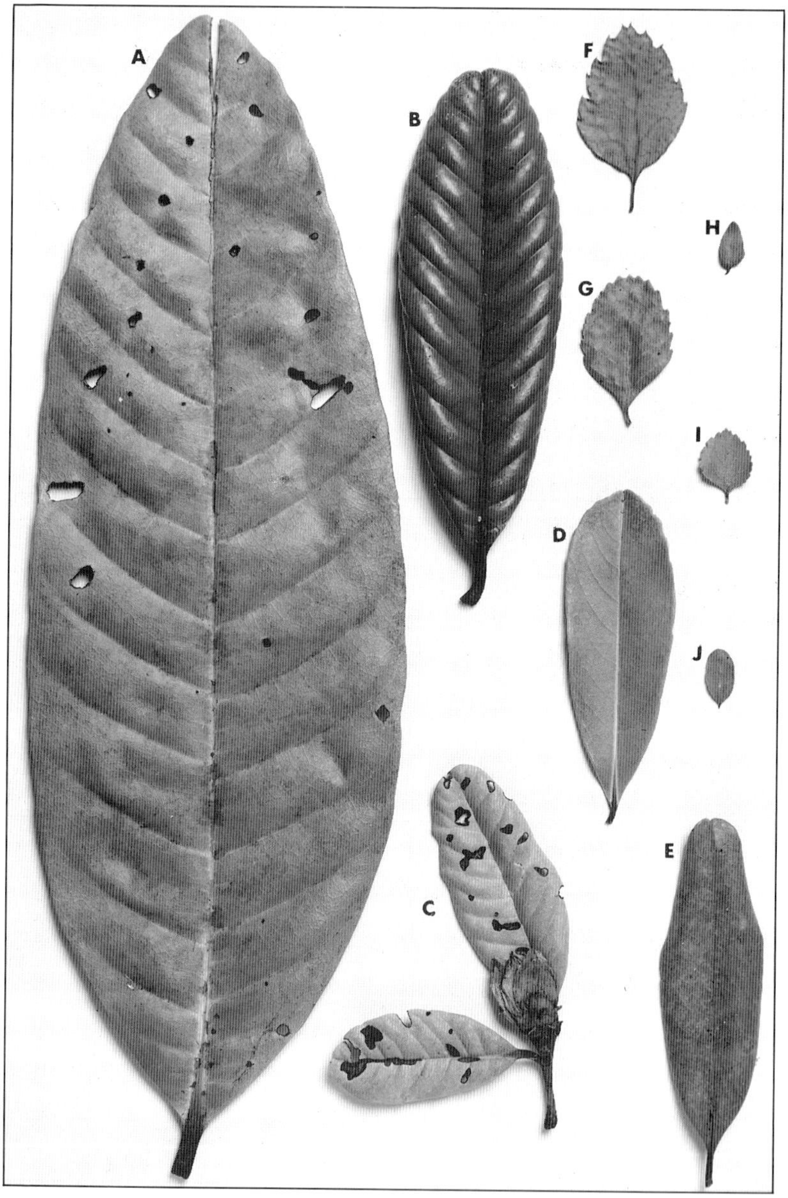


*Figure 70* New Zealand and New Caledonian *Nothofagus* leaves.
New Caledonia: *Nothofagus codanandra*; A, juvenile leaf; B, adult leaf; C, twig with a cupule with 3 flat seeds; D, *Nothofagus sp.*; E, *Nothofagus equilateralis*.
New Zealand: F, *Nothofagus fusca* (red beech); G, *N. truncata* (hard beech); H, *N. solandri* var. *cliffortioides* (mountain beech); I, *N. menziesii* (silver beech); J, *N. s.* var. *solandri* (black beech).
Photo: M. D. King.

*Figure 71* *Nothofagus* forest in the Travers Valley, northern South Island. Note the even treeline. Photo: J. W. Dawson.

## Structure and Composition of New Zealand Beech Forests

Seen from either the inside or the outside, pure beech forest in New Zealand has a very different appearance from conifer broadleaf forest. Externally (Fig. 71), for example when viewed on the steep side of a mountain valley, beech forest has a uniform, generally dark green colour and a smooth to moderately uneven surface. There are no emergent trees. There is none of the tangled profusion of the conifer broadleaf

forest within the beech forest (Fig. 72), particularly when rainfall is moderate or low. There are clearly fewer species present and because of the sparse undergrowth and absence of vines, one can walk freely among the rough-barked trunks. Vascular epiphytes too are generally absent, although there are often epiphytic mosses and lichens on the trunks and branches. The forest floor has a deep carpet of leaf litter, soft cushions of the pale green milk moss (*Leucobryum candidum*) and masses of the translucent fans of the kidney fern (*Trichomanes reniforme*). Among the scattered shrubs which may be present, the small leaved *Styphelia (Cyathodes) fasciculata, S. juniperina* (soft and prickly mingi-mingi respectively), *Coprosma rhamnoides* and *C. microcarpa* are most frequently encountered. Light filtering through the attractive layers of the beech foliage contributes to the typically tranquil, orderly atmosphere of this type of forest.

Beech forest becomes more complex with higher rainfall, as on the west of mountains at higher altitudes. Its growth appears more luxuriant, largely as a result of the profusion of lichens, mosses, liverworts and moss-like filmy ferns clothing the forest floor and trunks and branches of the trees. Where the canopy is partly open there are also more species of small trees and shrubs involved although none is restricted to beech forests. Broadleaf (*Griselinia littoralis*), *Pseudopanax simplex*, *Coprosma foetidissima* and others may contribute to a subcanopy and there is often a shrub layer of small-leaved coprosmas, the small-leaved *Myrsine divaricata*, the mountain horopito (*Pseudowintera colorata*) and others.

Larger ferns on the forest floor include the well-known crape fern or Prince of Wales Feathers (*Leptopteris superba)* and *Blechnum discolor.* Two small flowering plants growing in moss cushions are characteristic of this type of forest — *Luzuriaga parviflora* with spreading wiry stems, two-ranked leaves and large spongy white berries, and *Libertia pulchella* with half erect fans of grasslike leaves.

These wet beech forests in New Zealand are sometimes swathed in mist; forests of a similar character elsewhere in the world are often referred to as cloud or mossy forest, or more romantically 'goblin forest' or 'elfin woodland'.

At the other extreme, some of the beech forests in relatively dry sites on the east of the South Island mountain axis are among the simplest of forests anywhere, consisting of virtually a single species — mountain beech (Fig. 111).

*Figure 72 (opposite)* Interior of red beech (*Nothofagus fusca*) and silver beech (*N. menziesii*) forest near the Maruia Saddle, South Island. Photo: J. H. Johns.

As a generalisation, where there is a trend away from mild, moist conditions and fertile soils towards cooler or drier conditions and less fertile soils, or combinations of these, there is often a transition from conifer broadleaf to beech forest.

## Lowland Beech Forest

Black beech and hard beech are often referred to as the lowland beeches because they are generally found in the same altitudinal zone as the conifer broadleaf forest. They reach their southern limits in the northern South Island, with a disjunct occurrence of hard beech in south Westland.

The lowland beeches generally do not become components of the conifer broadleaf forest, however, but form instead mainly narrow strips of forest on the thin, stony, infertile soils of ridge crests (Fig. 73) or cliff edges. Often the change from dense conifer broadleaf forest to open beech forest is very abrupt. In the steepest, stoniest and driest sites only black beech may be present[94] but elsewhere the two species may intermingle or, on somewhat better soils, hard beech may predominate.

## Montane Beech Forest

In many places in New Zealand, conifer broadleaf forest gives way at higher elevations to a beech forest dominated by one or more of the montane beeches — red, silver or mountain. This occurs because of the decrease of temperature as altitude increases. Naturally the altitudes at which the transition takes place varies with latitude: at 38 to 39°S from East Cape to the central North Island it ranges from about 850 to 1000 m; at 41°S in the southern North Island it is at about 750 m; at 45°S in the south-west South Island it is at about 450 m; and at the southern end of the South Island at 46°S, montane beech forest approaches sea level in places.

The three montane beech species differ ecologically. Red beech gives way to silver beech, which gives way to mountain beech as temperatures become colder and soils become thinner and less fertile. Thus if the three occur together on the same mountain, red beech often dominates immediately above the conifer broadleaf forest giving way upwards to silver beech, which in turn gives way to mountain beech towards the treeline.

*Figure 73* Aerial view of inland Taranaki forests with strips of dark-coloured lowland beech forest on the ridges and conifer broadleaf forest on the slopes and in the valleys. Photo: White's Aviation.

Mountain beech is also more tolerant of drier conditions than the other two species so on the drier eastern sides of the mountain ranges, particularly in the South Island, it is often the only beech species present or even virtually the *only* tree species present as already mentioned. Conversely, on wetter sites, for example to the west of the mountain ranges, mountain beech becomes much less common or absent and its place at treeline may be taken by silver beech.

## Treeline

Where beech forest is present and when temperatures become too cold for tree growth as the altitude increases, there is often an abrupt change from the uppermost trees to a vegetation of alpine herbs and/or shrubs.

The highest treelines are those where mountain beech is present. In the central North Island at 39°S the mountain beech treelines range from about 1350 to 1500 m; in the north-eastern South Island at 42°S, 1300 to about 1550 m; and in the southern South Island at 45°S from 950 to 1250 m.

Compared with treelines at comparable latitudes on a continent, those in New Zealand are unusually low.[95,96] For example, at 39°N in the Rocky Mountains of North America, the treeline can be as high as 3630 m. This apparent anomaly seems to relate to the difference in summer temperatures between continents and islands at comparable latitudes. Land heats up much more in summer than the sea, so a continent will attain higher temperatures at given latitudes and altitudes than a narrow island surrounded by the relatively cooler sea. Consequently vegetation zone boundaries, including the treeline, are higher on continents than on islands. It is estimated that a mean warmest month temperature of 10°C determines the upper limit of tree growth and this seems to hold as true for the relatively low beech treelines in New Zealand as for the higher continental treelines.

On this basis it might be expected that mountains furthest from the sea in New Zealand with more 'continental' climates would have higher treelines. This proves to be the case. The highest treelines in New Zealand are on Mt Ruapehu in the central North Island (ca. 1500 m) and on the mountains at the centre of the northern South Island (ca. 1550 m). Inverted treelines occur where there are flat valley floors not far below the normal treeline. In frosty weather the coldest air sinks to the valley floor, which as a result experiences much lower temperatures than the valley sides, despite their higher altitude. In these circumstances the valley floor is occupied by an open vegetation of an alpine/subalpine character, while the slopes support forest.

## Beech Gaps

There are regions in New Zealand where beech species are completely absent, even though at appropriate altitudes conditions seem entirely suitable for them. Such regions are: Mt Taranaki (Egmont), an isolated volcanic cone in the west of the North Island; the southern Ruahines and adjacent northern Tararuas in the south of the North Island; in the South Island, a strip to the west of the high central third of the mountain axis (Westland) with a narrower gap on the east; and Stewart Island. For the central South Island at least it is suggested that beech was eliminated there during the last glaciation and because of its slow rate of dispersal has not yet fully reinvaded from southern and northern refugia.[97] As a consequence more readily dispersing conifer broadleaf forest species have established in sites perhaps suited to beech species, particularly at higher elevations. (See Chapter 12 for further discussion of the discontinuous distributions (disjunctions) of the beeches and other native species.)

## Intermediate Forest Types

Forests comprising mixtures of conifer broadleaf forest species (particularly the conifers) and beech can be found throughout New Zealand. In the north of the North Island there are even a few localities where hard beech intermingles with kauri (*Agathis*).

Such intermediate forests may be found in the transition from conifer broadleaf forest to pure beech forest on mountain slopes, but to the north and south of the Westland beech gap, there are more extensive mixed forests at lower and middle elevations. Some workers have suggested that this results from the slow replacement by advancing beech forest of the Westland conifer broadleaf forest.[69,97,98]

## Beech Forest Regeneration

The species of southern beech are all relatively light-demanding. Seedlings may establish on better-lit parts of the forest floor but then 'stand still' until more light is available from the death of a canopy tree or as a result of more extensive openings caused by violent gales. Such openings or clearings may soon be filled by a dense 'crop' of vigorous young beech trees.

Where beech forest is destroyed by fire followed by the establishment of manuka and/or kanuka, then there is sufficient light at ground level for beech to become established, grow strongly and, in time, to overtop and shade out the pioneers.

### Relationships of the New Zealand Beech Forests

Unlike the New Zealand conifer broadleaf forest, beech forest shows few similarities with tropical rain forest. Indeed its simpler structure, fewer species and general lack of vascular epiphytes and lianes, linked with the relationship of the dominant trees with true beech of the northern hemisphere, might indicate a vegetational affinity with north temperate deciduous forest.

New Zealand beech forest, however, differs from north temperate deciduous forest in certain respects. The leaves of the trees are evergreen and relatively small. The forest generally occupies infertile sites and forms a slowly decomposing, low-nutrient leaf litter which, because of its acidity, accelerates leaching and further reduces soil fertility. North temperate deciduous forest, on the other hand, generally occupies more fertile soils and forms a rapidly decomposing leaf litter, high in nutrients, which tends to improve the fertility of the soil.

In the respects in which New Zealand and other southern hemisphere beech forests differ from north temperate deciduous forest, there is a fair measure of agreement with the vast coniferous forests of high northern latitudes. This has led to the suggestion that these two forest types of cooler infertile sites in the two hemispheres are approximate ecological equivalents.[99]

However, as there are a number of deciduous species of *Nothofagus* in South America and one in Tasmania, it might be truer to say that *Nothofagus* forest lies somewhere between the north temperate deciduous and coniferous forests. Within New Zealand, in terms of nutrient demand and leaf size, red beech would come closest to the former and mountain beech to the latter.

## CHAPTER 6

# The Small Leaved Shrub Problem

If the profusion of vines and epiphytes in the conifer broadleaf forests is the first thing that strikes the overseas botanist, the second is probably the prevalence, in various open habitats and some forest types, of densely twiggy shrubs with very small leaves and slender, wiry stems. Cockayne called them 'divaricating shrubs' because of the wide or divaricate branching angles (to 90° or more) of the twigs.[100] Another feature of many divaricates is the vigorous growth of lateral twigs from all, or almost all, of the leaf angle buds of each branch, even those nearest its tip. In 'normal' shrubs the growing tip of a branch grows strongly and forms a chemical inhibitor which prevents the lateral buds for some distance below from themselves growing into branches. With many divaricate shrubs, growth of a new branch, although vigorous at first, eventually becomes weak and as a result the no-longer-inhibited lateral buds grow out strongly until their activity too declines. It could almost be said that the branching of these shrubs is uncontrolled in comparison with more familiar growth patterns. The result of this distinctive branching behaviour is a densely interlaced, springy, more or less ball-like shrub. Where these grow in windswept situations on hill tops or at the coast they form low hummocks which are so dense that a person can stand on them (Fig. 74), or even jump up and down on them, without breaking through the surface. At first acquaintance all divaricate shrubs tend to look the same, but, according to Greenwood and Atkinson,[62] there are in New Zealand 54 species of this form (including the divaricating juveniles of some trees) belonging to 20 genera and 17 families (Figs. 75, 76, 77). A further 9 species are illustrated in Eagle's *Trees and Shrubs of New Zealand,* but have not yet been named formally. Nine species from an additional 6 genera (including *Nothofagus, Neomyrtus, Lophomyrtus* and *Teucridium)* and 4 families are termed semi-divaricating. Some of these have small leaves and slender twigs but their branching angles are relatively narrow.

Some divaricate shrubs are the persistent juvenile stages of forest trees, which undergo a dramatic change in branching pattern and an equally dramatic increase in leaf size when they change to the adult state (Figs. 78, 79).

Two species have divaricate juveniles with leaves which are remarkably variable from long and narrow to short and round and from deeply lobed and dissected to entire — pokaka (*Elaeocarpus hookerianus*) (Fig. 78) and *Pittosporum turneri*. The non-divaricating vine *Parsonsia heterophylla* has similar juvenile leaves. Divaricating shrubs may themselves have juvenile forms. *Melicope simplex* and *Pseudopanax anomalus* have simple adult leaves but compound juvenile leaves. Several divaricate pittosporums have strongly dissected or narrowly elongate juvenile leaves. Perhaps most remarkable of all, *Aristotelia fruticosa* has a juvenile which is not divaricate, but has a similar range of leaf forms to that of pokaka. Cockayne[101] says: 'The various forms assumed by the (juvenile) leaves of *A. fruticosa* are almost beyond belief.'

Little is yet known about the precise habits of growth of most

*Figure 74 (opposite)* A remarkable windswept shrub association comprising a number of species of divaricate shrubs and juveniles as well as some vines. Note the figure standing on the surface of the shrubbery. Mt. Kaukau near Wellington, southern North Island. Photo: J. W. Dawson.

*Figure 75* Habit of *Pittosporum obcordatum* in cultivation at Otari Gardens, Wellington. Photo: J. W. Dawson.

divaricates, but it is clear that they are not uniform, for example:

(a) in some species the interlacing tendency of the twigs is enhanced by the zig-zag pattern of the internodes;

(b) in others some of the shoots remain very short and bear most of the leaves of the shrubs;

(c) in those species where there is more than one bud in the leaf axils, development of some of the additional buds after a period of dormancy further complicates an already complicated branching pattern.

There are differences of opinion about whether divaricating shrubs which are exactly equivalent to those of New Zealand occur elsewhere in the world,[62] but it seems that only in New Zealand are they so prevalent in both the flora and the vegetation.[102]

The majority of New Zealand divaricates belong to genera in which there are also local forest species of normal branching patterns, usually with much larger leaves. In the following list the numbers of divaricate and non-divaricate species are given for each genus.

*Figure 76 Pseudopanax anomalus.* Although seedlings have small palmately-compound leaves, adult leaves are simple and 1-2 cm long. Other New Zealand species of *Pseudopanax,* such as five-finger (*P. arboreus*) at top right and *P. laetum,* have large, palmately-compound leaves with leaflets up to 25 cm long. Photo: J. W. Dawson.

*Figure 78* Pokaka (*Elaeocarpus hookerianus*). Juvenile divaricate form above, adult form below. Photo: J. E. Casey.

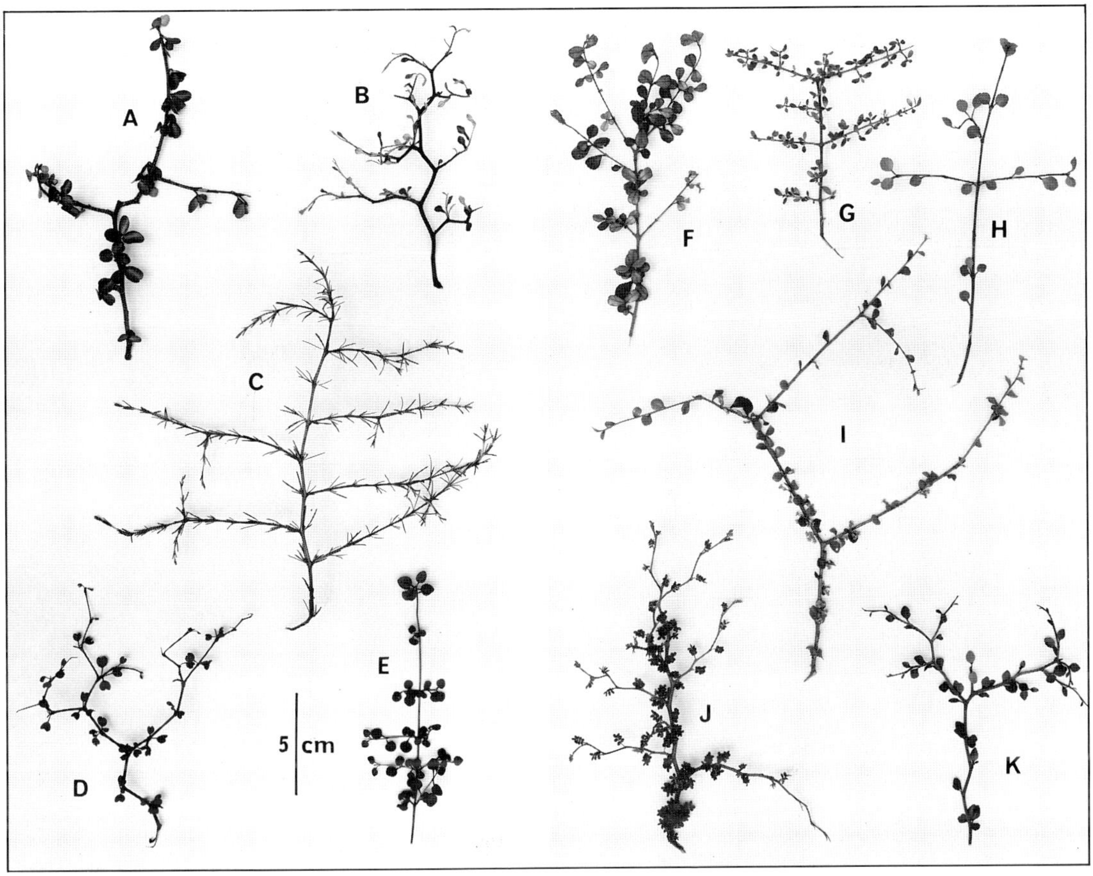


*Figure 77* Leafy twigs of some of the New Zealand divaricate species, with one exception each from a different family:
A, *Myrsine divaricata*.
B, *Corokia cotoneaster*.
C, *Coprosma acerosa*.
D, *Muehlenbeckia astonii*.
E, *Teucridium parvifolium*.
F, *Lophomyrtus obcordata*.
G, *Coprosma wallii*.
H, *Melicope simplex*.
I, *Pittosporum obcordatum*.
J, *Sophora prostrata*.
K, *Pseudopanax anomalus*.
Photo: J. E. Casey.

| | *Divaricate* | *Non-divaricate* | *Divaricate juvenile* |
|---|---|---|---|
| *Violaceae* | | | |
| Melicytus (incl. Hymenanthera) | 5* | 8* | |
| *Polygonaceae* | | | |
| Muehlenbeckia | 1 | 4 | |
| *Pittosporaceae* | | | |
| Pittosporum | 5 | 14 | 1 |
| *Elaeocarpaceae* | | | |
| Elaeocarpus | | 1 | 1 |
| Aristotelia | 1 | 1 | |
| *Malvaceae* | | | |
| Plagianthus | 1 | | 1 |
| Hoheria | | 4* | 2 |
| *Escalloniaceae* | | | |
| Carpodetus | | | 1 |
| *Leguminosae* | | | |
| Sophora | 1 | 1 | 1 |
| *Moraceae* | | | |
| Streblus (Paratrophis) | | 2 | 1 |
| *Icacinaceae* | | | |
| Pennantia | | 1 | 1 |
| *Rhamnaceae* | | | |
| Discaria | 1 | | |
| *Rutaceae* | | | |
| Melicope | 1 | 1 | |
| *Araliaceae* | | | |
| Pseudopanax | 1 | 13 | |
| *Cornaceae* | | | |
| Corokia | 1 | 2 | |
| *Myrsinaceae* | | | |
| Myrsine | 1 | 6 | |
| *Rubiaceae* | | | |
| Coprosma | 28* | 21 | |
| *Compositae* | | | |
| Olearia | 7 | 24 | |
| *Podocarpaceae* | | | |
| Prumnopitys | | 1 | 1 |

* includes some species not yet named, but illustrated in Eagle

Greenwood and Atkinson[62] consider that 26, or nearly half, of the divaricate species they considered are largely restricted to one vegetation type: forest (10); forest margins (3); scrubland[104] (5); shrubland[104] (6); grassland and other open vegetation (2). The other 28 species can be found in two or more vegetation types. Some of these are remarkable in their range. For example *Coprosma rhamnoides* at one extreme grows in shade and shelter on open forest floors and there forms attractive shrubs with leaves which tend to be arranged in horizontal layers. At the other extreme it may be found in the open on hill tops exposed to violent gales, where it forms or contributes to dense shrub hummocks shaped and channelled by the wind (Fig. 74). The five-finger relative *Pseudopanax anomalus* is a similar case.

Within the genera concerned, the 'normal' species differ so markedly in appearance from the divaricates that it is difficult to believe that they are related. Apart from the differences in branching pattern the leaves of the normal species are often hundreds of times the area of those of the divaricates. Even more surprising is the fact that natural hybrids are not uncommon between divaricate and non-divaricate species of several genera — *Melicytus, Aristotelia, Sophora, Melicope, Pseudopanax, Corokia, Pennantia* (Fig. 79), *Myrsine, Coprosma*[105] (Fig. 80) (3 pairs of species) and *Olearia*. In some cases the first generation hybrids are partly fertile and give rise to bewildering arrays of second generation hybrids known as 'hybrid swarms'.

It is generally believed that the divaricates are derived and specialised forms and the fact that a number of them are able to hybridise with large-leaved relatives might indicate that this distinctive growth habit evolved in relatively recent geological times.

We now come to the problem of how and why divaricate plants became such a conspicuous feature of the New Zealand flora. If there were only one species of this form it might be dismissed as a chance aberration, but as there are many, and these are mostly unrelated to one another, some special environmental factor or combination of factors selecting for the habit is indicated. A contrary view is that of Went,[106] who suggests a virus-like transmission of chromosome segments, bearing divarication inducing factors, from species to species, genus to genus and family to family. Because of the difficulty of envisaging a mechanism for such transfers this idea has not received much support, although it does not seem so outlandish in these days of genetic engineering as it would have done some years ago.

*Figure 79* Hybridism in *Pennantia*. At upper left is the small-leaved divaricate juvenile of *Pennantia corymbosa* (kaikomako) and at lower left a portion of the non-divaricate larger-leaved adult. Kaikomako ranges throughout the country and is a small tree mostly found at forest margins. On the right is a large, thick leaf of *Pennantia baylisiana*. This, the only other species of *Pennantia* in New Zealand, is restricted to the Three Kings Islands and is represented there by a single small tree. Like kaikomako, *P. baylisiana* is dioecious and the sole tree is a female. It has been reproduced by cuttings and a flowering specimen is now established at Otari Plant Museum in Wellington. It was noticed that some berries were formed by this tree and some of the seeds in these germinated to form several identical plants now several years old, although not yet flowering. It seems likely that the female flowers of *P. baylisiana* were pollinated by male plants of kaikomako in the vicinity and that the progeny are hence probably hybrids between the two species. (An alternative possibility in that the suggested hybrids are juveniles of *Pennantia baylisiana* derived from embryos without fertilisation (apomixis).) Foliage from the suggested hybrid is at the centre of the photo. Its leaves are dark green and shiny like those of *P. baylisiana*, but are much smaller and toothed like those of kaikomako. There is no clearly marked juvenile form in these plants. There are several other cases of hybrids between morphologically very different species in New Zealand — *Coprosma robusta* x *C. propinqua* (Fig. 80), *Melicope ternata* x *M. simplex*, *Corokia buddleoides* x *C. cotoneaster* and a number of others.
Photo: J. E. Casey.

*Figure 80* Hybridism in *Coprosma*. Left: a leaf of karamu (*Coprosma robusta*); right: twigs of *Coprosma propinqua* showing the very small leaves; centre: the natural hybrid between these two species. This hybrid was originally described as a species — *C. cunninghamii*. Allan[100] proved that it was a hybrid by artificially crossing the two species.
Photo: J. W. Dawson.

## Climatic Theories

However, most theories about divaricating shrubs regard them as an example of parallel evolution in response to some environmental factor, but there is little agreement yet on what that factor is. The first to speculate late last century was Diels,[107] who as a young man wrote a remarkable thesis on New Zealand vegetation even though at that stage he had not seen it. He based his account on herbarium and living plants available to him in Berlin, earlier publications and information from correspondents including Cockayne. Diels concluded that the divaricate habit was an adaptation to drought and suggested that shrubs with this form had evolved from forest ancestors at a time when the South Island axial ranges were higher than they are now and the rain shadow effect to the east of them, accentuated by drying winds, was much more extreme. Diels also suggested that coldness during glacials was a contributing factor in the evolution of this growth form, and Rattenbury[9] later suggested that the influence of coldness could have been indirect by inducing physiological drought of the root systems. Cockayne[100] accepted and extended Diels' interpretation, paying particular attention to trees with divaricating juveniles. He saw these as being ultimately derived from pre-ice age tree ancestors which completely lacked divaricate branching at any stage. During the cold and dry conditions in the east during glacials divaricating shrubs evolved from the trees and with the return of warmer, moister conditions some of these regained their tree form, although only after passing through a small-leaved, divaricating juvenile phase. Cockayne regarded the last growth pattern as an example of 'ontogeny recapitulating phylogeny', that is the persistance of ancestral traits in the embryonic or, in this case, juvenile phase of a life cycle. In support of this evolutionary sequence he noted that ribbonwood (*Plagianthus regius*) has a divaricating juvenile on the main islands but not on the outlying Chatham Islands. He regarded the population on the latter group as the ancestral form, which survived in the milder, oceanic climate of a small island.

Wardle[108] took a different view of trees with divaricating juveniles. Noting that divaricates generally are conspicuous in 'dryish' eastern forests, he suggested that divaricate juveniles might relate to existing climates which tend to periodic drought and that the changeover to the larger-leaved adult form takes place once a deep, efficient root system has developed. At times of greater aridity, perhaps during glacials, some divaricate juveniles would become 'fixed' flowering and fruiting divaricate shrubs. Thus according to Cockayne the sequence would have been: non-divaricating tree — divaricating shrub — tree with divaricating juvenile; and according to Wardle: non-divaricating tree — tree with divaricating juvenile — divaricating shrub.

More recently, Wardle's hypothesis concerning microphyllous juveniles has been applied on Reunion and Mauritius Islands in the Indian Ocean where microphyllous, although not divaricating, juveniles are unusually common on the drier sides of the islands.[26]

With regard to *Sophora,* Cockayne held a view in conflict with his own theory but in accord with that of Wardle: the divaricate shrub *Sophora prostrata* is interpreted as a flowering and fruiting fixed juvenile of *S. microphylla*. Godley questions this view,[109] pointing out that the leaves and stems of *S. prostrata* and the divaricate juvenile of *S. microphylla* are quite dissimilar, as are the flowers and seed pods of the two species. He proposes an interesting and quite different explanation — the extremely variable *Sophora microphylla* (with everything from non-divaricate to strongly divaricate juveniles depending on locality) may derive from an ancient hybridisation between the non-divaricate and relatively large-leaved small tree *S. tetraptera* and the dwarf divaricate shrub *S. prostrata*. He further suggests that all trees with divaricate juveniles may have originated in a similar way.

McGlone and Webb,[110] in response to the 'moa theory', which we will consider later, reaffirmed and refined the climatic explanation of the divaricating habit. They believe 'that the divaricating habit is an adaptation which enables the plant to resist damage from wind, frost and desiccation, while retaining enough flexibility to exploit a wide range of habitats'. They relate this growth habit to what they regard as New Zealand's distinctive climate: 'as a consequence of its position in the mid-latitudes and its isolation in a large area of ocean, New Zealand has a mild, generally humid, windy climate. However, it is also a very variable climate and the continental pattern of a cold winter and a consistently warm summer does not apply here.'

They point out that in many forest areas in New Zealand, night frosts may occur at any time of the year, often to be followed by quite warm days. Also in areas to the east of the mountain ranges, dry often very warm föhn winds may be frequent in spring and summer and are sometimes followed by cool or even frosty weather especially inland. In such circumstances deciduousness would seem an inappropriate adaptation, but McGlone and Webb believe that the mostly evergreen[111] small-leaved divaricating shrub is well suited to survive the inclement and to take advantage of the favourable spells of weather. They see the ball-like divaricating shrub, with its internal as well as external twigs, growing points and leaves, serving as a wind screen, frost screen and heat trap. With regard to wind they see three benefits for the internal leaves, which outweigh the disadvantage of their being partially shaded: the drying effect of the wind is reduced; abrasion by wind-borne particles is reduced; and the tendency of the interlaced mass of twigs to move as a whole means that twigs and leaves are less likely to hit each other and become damaged. With frost they have observed in a number of divaricating species that 'even in severe frosts in which exposed leaves are frozen the interior leaves of the shrubs are unaffected'. Finally, concerning the heat trap idea, they say, 'besides possibly acting as a frost screen, the protective network of branches may, on cold, but sunny days, act as a heat trap, raising the temperature of the air mass inside the shrub, and this may permit higher rates of photosynthesis'. With divaricate juveniles McGlone and Webb agree with Wardle that the changeover to the adult form may follow the development of a deep root system, but also suggest that the changeover may take place once a height above ground frost level is attained.

They suggest that the genera in New Zealand which have some divaricating species probably originated under at least subtropical conditions and had the opportunity to evolve cold-tolerant forms because New Zealand's isolation and lack of mountains during the Tertiary largely precluded the establishment of genera which had evolved in colder climates: 'we believe that the divaricating plant habit was one of the responses of this essentially subtropical flora to the onset of glacial climates.'

Indeed, evidence of trends in branching patterns and leaf sizes in tropical and subtropical forest genera in relation to decreasing temperature does indicate that the divaricating habit might represent the end point of such trends. In places in the world such as coastal Brazil,[112] where rain forest extends from the equator to subtropical latitudes, a correlation has been observed between decreasing temperatures with increasing distance from the equator and a replacement of big-leaved, sparsely branched species with progressively smaller-leaved and more freely branched species. New Zealand's isolation when climates became colder would have enabled such a trend to continue to its ultimate extreme.

### Moa Theory

We come now to the 'moa[113] theory' elaborated by Greenwood and Atkinson,[62] which has found favour with some but disbelief from others. Some of the negative reactions, I think, stem from the fact that the moas are extinct. The 'moa theory' has a science fiction ring to it, just as would a 'dodo theory' or, to extend into myth, a 'dragon theory'. At our present state of knowledge, however, the moa theory merits serious consideration along with the climatic theories.

Essentially, Greenwood and Atkinson propose that the divaricating shrub habit is probably unique to New Zealand and results from an environmental factor that was also unique to New Zealand, namely the moas. Species of similar appearance to divaricating shrubs elsewhere in the world either have relatively large leaves and/or are spiny. Spininess of shrubs is considered to be a defence against the browsing of soft-nosed mammals, but it would be ineffective against large, browsing birds with their hard beaks. With the exception of matagouri *(Discaria toumatou)* New Zealand divaricates are not spiny, although there is a tendency that way with some reduced shrubs in exposed habitats. Matagouri has spiny relatives in Australia and South America so probably evolved its spininess outside New Zealand.

Greenwood and Atkinson considered the essential features of divaricating shrubs to be: interlaced branching; small leaf size; stem toughness; and leaves which are smaller on outer than on inner branches. In their original publication they assumed that moas could not bite with a cutting action, but clamped, pulled and broke off portions of foliage. This would have made it difficult for them to browse on divaricates as the twigs are wiry and difficult to pull off and, when broken, difficult to disentangle. It was further suggested that where the outermost twigs had very reduced leaves and a dead appearance they would be rejected as unpalatable. With regard to trees with divaricating juveniles, they suggest that the change to adult branching and foliage took place above the reach of moas.

Later work showed that the assumption about moa eating habits was incorrect. Burrows[114] examined the plant material contained in fossilised moa gizzards and found that it mostly 'consisted of twigs, commonly 1.5–6.0 mm thick and 10–30 mm long. Some of these twigs are of *Olearia virgata* and *Plagianthus regius* (both divaricating species), and much material appears to have come from *Coprosma,* a genus which contains a large number of divaricating species. Most of the twigs were cleanly cut, probably by the shearing action of the moa beak, rather than broken'.

Greenwood and Atkinson[115] and Lowry[116] suggested that although the divaricate habit may not have prevented moa browsing, the many growing points separated by the wide branching angles would have enabled plants of this form to survive and quickly recover from browsing.

Greenwood and Atkinson note that divaricates are most abundant on fertile soils such as those of river flats and suggest that this is because plants on such sites would be more nutritious and therefore favoured by moas. On the other hand they point out that there are few divaricates on outlying islands not reached by moas and none on cliffs or as epiphytes on branches: 'moas would have experienced difficulty in browsing either on cliffs or up trees!'

In opposition to this view, McGlone and Webb[110] suggest that the prevalence of divaricates on river flats and similar places derives from the fact that such sites are particularly frosty. They also ask why divaricates, if they were maintained on river flats by browsing, have not been replaced in the centuries since moa extinction by faster growing, larger-leaved shrubs and trees.

To throw yet another hat into the ring, river flats are also subjected to floods from time to time, so perhaps the divaricate habit is flood-resistant and in trees with divaricate juveniles the change to the adult state takes place above flood level? It is perhaps relevant here that McGlone and Webb believe that river beds, dunes and wind-induced scrub were probably the only places in the lowlands during the warmest phases of the interglacials, which could serve as refugia for divaricates. The 'flood' hypothesis undoubtedly raises a number of questions, such as why there are no divaricates within the flood zone of river cliffs, but it is not alone in this respect. Perhaps divaricate plants are tolerant of a number of environmental factors rather than just one.

The one thing that is clear in all of this is that the divaricating shrub story has not yet been fully told and that we need to know much more about the morphological, physiological and ecological details of the plants concerned.

Finally, not all small leaved shrubs in New Zealand are of divaricate form. The large genus *Hebe* has many scale-leaved 'whipcord' species, mostly found at higher altitudes, which are probably derived forms, as are the many needle-leaved species of *Dracophyllum* and some small-leaved but not divaricating species of *Olearia, Coprosma* and other genera. We need to keep these in mind when theorising about shrubs which are divaricating as well as small-leaved.

## CHAPTER 7

# Open Habitats in the Lowlands

At the time of European settlement some lowland areas, particularly in the North Island, were occupied by bracken fern or shrubland dominated by manuka (*Leptospermum scoparium*) and/or kanuka (*Kunzea ericoides*). It is now believed that these areas were originally largely under forest, which was destroyed and prevented from returning by repeated fires of both Maori and natural origin. Many of the fires caused by the Maori were accidental, no doubt, but others were deliberate to encourage the growth of bracken fern whose rhizomes were an important source of starch.

Much of the eastern side of the South Island, as well as the Southland Plain, supported native grassland: short tussock grassland on the driest sites (the Marlborough Plains and adjacent hill country, the Canterbury Plains and foothills, and the Central Otago hill country) and tall tussock grassland of red tussock (*Chionochloa rubra*) on the moister Southland plains and eastern Otago hills. For some time it was assumed that the short tussock grassland was the original vegetation cover; the regions in which it predominated seeming too dry to support forest. When exotic trees were planted and thrived, there must have been some doubt about this view, but it was not until relatively recent times that evidence came to light indicating that these areas of short tussock grassland originally supported forests and forest/tussock shrubland mosaics. This evidence, to quote Molloy,[117] includes 'surface logs and forest wind throw hummocks and hollows in now tree-less areas; buried wood, charcoal, and other plant remains; buried (forest) soil profiles; and a number of relict plants and soils'. It is suggested that forests, particularly in the east of the country, became more susceptible to fire following a change to more variable climates with periodic droughts after about 1800 years ago.[118]

Identification of wood fragments has allowed a generalised reconstruction of these former Canterbury forests.[119] On the western ranges bordering the plains, beech forest extended in most places from tree line down to about 1000 m altitude. Below this, down to about 400 m, came a mixed beech-podocarp forest, and on the plains themselves from 300 m to sea level, there was a type of conifer broadleaf forest dominated by matai (*Prumnopitys taxifolia*) and totara (*Podocarpus totara*). Near the sea on swampy sites were stands of kahikatea (*Dacrycarpus dacrydioides*) forest, of which Riccarton Bush is a surviving example. Older, shallow soils were occupied by low forests of kanuka (*Kunzea ericoides*), which may, in part at least, have resulted from natural fires. Following the extensive destruction of the forests by fire during the period when the Maori was hunting moas in Canterbury, the higher altitude forests on the inland ranges were replaced by shrubland and narrow-leaved snow tussock (*Chionochloa rigida*), which moved down from higher altitudes, and red tussock (*C. rubra*) which moved out from poorly drained sites on to sometimes quite well-drained hillsides. On the plains themselves, short tussock grassland became established — 'The species

*Figure 81* Short tussock grassland dominated by fescue-tussock (*Festuca novae-zelandiae*) on the Mt. Hay Station, Tekapo, central South Island. Photo: E. J. Godley.

forming this grassland would have come largely from river and stream bed floras, from small enclaves of tussock within the forest, from the florulas of rock outcrops, and from among the species with short life-cycles from the alpine snow tussock above.'[120]

Short tussock grassland (Figs. 81, 82) is characterised by the light brown tufts, up to about half a metre high, of hard tussock (*Festuca novae-zelandiae*) together with silver tussock (*Poa cita* = *P. caespitosa*), blue tussock (*Poa colensoi*) and blue wheat grass (*Elymus rectisetus*). The most

*Figure 82* Short tussock grassland mostly of *Festuca novae-zelandiae* on hills north of the upper Waitaki River. Drought resistant shrubs grow on rock outcrops and along the bottoms of gullies where more moisture is available. Photo: J. W. Dawson.

*Figure 83* Matagouri or 'Wild Irishman' (*Discaria toumatou*) showing spines and flowers. Photo: J. W. Dawson.

conspicuous shrub, particularly on stony, moister sites, is matagouri or 'wild Irishman' (*Discaria toumatou*) (Fig. 83). Matagouri is freely branched and interlaced and set with sharp spines that make it very unpleasant to push through. Small leaves are produced in the spring, but the shrubs are quite leafless during the summer. Two leafless brooms, *Carmichaelia petriei* and *C. robusta*, with slender switch-like branches, may stand above the tussocks in drier sites, while among smaller shrubs are *Pimelea oreophila* and *Styphelia nesophila* (= *Cyathodes fraseri*). Notable among the larger herbs is the spaniard *Aciphylla subflabellata,* whose close rosettes of much-divided leaves look like clusters of grey, slender, sharply-pointed knitting needles (Fig. 84) and whose equally spiny bracts form cage-like enclosures for the seeds (Fig. 85). The golden spaniard (*A. aurea*) also often descends into short tussock grassland from higher altitudes.

One of the driest regions supporting short tussock grassland was the Central Otago hill country and here overgrazing with sheep and the later depredations of rabbits reduced the landscape to virtual desert (Fig. 86). The tussocks were largely replaced by grey patches of *Raoulia australis,* unflatteringly termed 'scabweed'.

Let us return to the red tussock grassland at the moister southern end of the South Island. A recent study[121] of fossil pollen from bogs built up over the past 12000 years on the Longwood Range in Southland, indicates that forests also covered the southern end of New Zealand until about 1000 years ago. The species represented suggest that kahikatea (*Dacrycarpus dacrydioides*)-dominated forest occupied moister and swampy sites, rimu (*Dacrydium cupressinum*)-dominated forest drier sites and silver beech (*Nothofagus menziesii*) forest the higher altitudes.

After 1000 B.P. (before present) the proportion of forest pollen decreased and spores of bracken fern and grass pollen increased. After 700 B.P. there was a striking upsurge of bracken spores followed by a

*Figure 84 (left) Aciphylla subflabellata* with male inflorescences.
Photo: J. W. Dawson.

*Figure 85 (above) Aciphylla subflabellata*. Seedhead with most of the bract segments aligned vertically to form a cage around the seeds.
Photo: J. W. Dawson.

peak of grass pollen: 'We conclude that localised burning took place shortly after the arrival of Polynesians in Southland, which was followed some 200 years later by wholesale destruction of lowland forest by fire. Immediately after the destruction of the forests fernland established, but, with continued burning, grassland began to spread . . . The first European settlers found the Southland Plains in shrubland, fern, and tussock, with indications in some areas that succession back to high forest had begun.' A similar sequence has been described from Lake Poukawa in the eastern North Island.[122]

Thus it seems that in their original state, before the increasing frequency of natural and human-caused fires over the past few thousand years, the lowlands of New Zealand had few sites not occupied by forests. In such open habitats as there were, establishment of trees would have been precluded by some localised environmental factor: periodic flooding and disturbance of parts of riverbeds normally above water level; excessive salinity in some coastal sites; excessive drainage and sparse soil of cliffs and bluffs in the drier eastern part of the country, and the infertile soils, containing varying concentrations of toxic metals (nickel, chrome, magnesium), derived from the ultramafic rocks more commonly known as 'serpentine'.

*Figure 86* A barren landscape near Bannockburn, Central Otago. The hillside in the foreground above the road has patches of 'scabweed' (*Raoulia australis*).
Photo: J. W. Dawson.

*Figure 87* The Rakaia River, a typical braided river of the Canterbury Plains.
Photo: V. C. Browne.

Lowland bogs and swamps would also have been treeless, but as these alter with time under the influence of plant cover and generally give way to forest, they have already been considered in Chapter 4.

## River Beds

These vary in their form quite markedly depending on the amount of eroded material — rock fragments, sand, silt — being brought down from the mountains. Where there is more material than the river can readily transport to the sea, the river bed is steadily raised. At the other extreme, where eroded material is slight and there is a sufficient fall to the sea, the river steadily cuts down into its bed. The best examples of the former type are the rivers traversing the Canterbury Plains (Fig. 87). These rivers have in fact built the Plains with the debris they have carried down from the Southern Alps. Their beds are sometimes a few kilometres wide and are described as 'braided' as they exhibit a network of channels, of which only a few actually carry water at any particular time. As each water-filled channel builds up to a higher level than adjacent dry channels it may take only a slight freshening of water flow to divert the river into a new course at a lower level. Such channel changes, and occasional floods which fill the river bed from bank to bank, put riverbed plants at a constant risk of being swept away. When a portion of a riverbed has been stable and dry for a time plants begin to establish. First come species of *Epilobium* (*E. brunnescens, E. melanocaulon* and *E. microphyllum*) and flat circular patches, up to a metre in diameter, of *Raoulia tenuicaulis*. As wind-blown silt accumulates, *Raoulia hookeri, R. australis* and other epilobiums follow and, if stable conditions continue, taller plants appear: short tussock grasses, carmichaelias, hebes, parahebes, matagouri (*Discaria toumatou*) and others.[123]

Rivers that are cutting down have much narrower beds which are largely under water most of the time. Such rivers are often confined by quite steep banks or, in gorges, by high cliffs. Plants grow mostly on the banks or cliffs, although some may establish on 'islands' in the bed itself. Rivers of this type are quite common in the North Island and some parts of the South Island.

The alluvial flood plains beyond the river banks often support communities of small trees. The species concerned are not confined to such sites, but some are most frequent there, including ribbonwood (*Plagianthus regius*), species of lacebark (*Hoheria*) of the family Malvaceae, and perhaps most notably the small trees that bear New Zealand's best known flower, the kowhais (*Sophora*). *Sophora microphylla* is found throughout and *S. tetraptera,* with larger leaflets, in the eastern North Island. The flowers, which are pollinated by birds, are golden yellow, pendent and more or less tubular in form.

On the rocky banks and lower parts of cliffs of the river bed itself plants are subjected to more frequent floods and are much smaller. Common among herbaceous plants are ferns, including several species of *Blechnum,* and flowering plants such as the small-leaved, creeping species of *Gunnera* and the two subshrubby species of *Jovellana,* a genus closely related to *Calceolaria*. Among shrubs several willow-leaved koromikos (*Hebe*) are prominent, for example *Hebe salicifolia* and varieties of *H. stricta* , as well as several species of the related *Parahebe* of which one is appropriately named *P. catarractae*.

A number of species of leafless broom (*Carmichaelia*) are often encountered in these riverbeds. They have slender but tough, flattened, more or less trailing stems with no leaves, or leaves for only part of the year. In the tropics there are often distinct communities of shrubs growing within the flood zone of streams.[124] Some of the species grow nowhere else and are 'streamlined' with long, narrow leaves and tough, slender stems to minimise the force of the water during floods. Probably none of the New Zealand species is restricted to river beds, but some show similar modifications to those in the tropics. The narrow-leaved hebes look remarkably like the unrelated *Metrosideros operculata* of New Caledonian stream beds and the carmichaelias with their modified stems seem well suited to avoiding damage from floods. The species of *Notospartium* and *Chordospartium*, related to *Carmichaelia,* of the north-eastern South Island, have 'slender, drooping, leafless branches' and may also grow near rivers.

This leads to the speculation that in warmer Tertiary times when open, lowland habitats were in short supply, *Hebe, Parahebe*, and *Carmichaelia* may have been largely confined to riverbeds, diversifying into other open habitats when forests retreated during glaciations.

## Cliffs and Rock Outcrops

Inland river cliffs above flood level in moister areas support a number of distinctive species mostly with drooping leaves or stems. The most striking and attractive of these, is the large-leaved fern *Blechnum 'capense',* which sometimes covers large areas to the exclusion of other plants. Other species conspicuous at some localities are various willow-leaved koromikos (*Hebe*), the robust pendent sedge *Machaerina sinclairii,* and, in better-lit places, the slender drooping stems of *Cordyline banksii.* Parataniwha (*Elatostema rugosum*) (Fig. 63) in northern parts of the country may cover with its attractive purplish foliage extensive areas of cliff, in the wettest and shadiest places.

Cliffs and rocky slopes along the coasts are populated by hardier species able to withstand gales and salt.[125] In the most exposed situations low growing succulent plants predominate, including the native ice plant *Disphyma australe,* the glasswort *Sarcocornia quinqueflora, Apium prostratum* and *Senecio lautus.* In the North Island particularly, prostrate shrubs of the shiny-leaved taupata (*Coprosma repens,* known as the 'looking glass plant' in California where it is cultivated) are often prominent.

On more sheltered aspects, so-called mountain flax (*Phormium cookianum*) may form a close cover, in association with *Hebe macroura* in the East Cape region, *Hebe stricta* in central New Zealand and, in the southern South Island and Stewart Island, *Hebe elliptica* and several ferns.

On the south Otago coast, species of two predominantly alpine genera are conspicuous — *Celmisia lindsayi* on solid rock faces and stony debris and *Anisotome lyallii* on debris. Other forms of *A. lyallii* occur in Fiordland and Stewart Island.

Also in Fiordland and Stewart Island are distinctive coastal shrubberies of 'daisy trees'. These may overhang the sea or be separated from it by the *Hebe elliptica/Phormium cookianum* association. In more sheltered situations, *Brachyglottis rotundifolia* predominates, with its large, round, leathery leaves giving way in exposed sites to *Olearia angustifolia,* a handsome shrub with quite large, white, purple-centred flowers.

The drier inland and coastal cliffs of Marlborough have their own distinctive flora. Notable is the genus *Pachystegia* (Compositae) endemic to Marlborough and North Canterbury. Until recently it was thought that there was only one species but a study in progress indicates that there could be as many as five.[126] They are all small, spreading shrubs with thick, leathery leaves, densely furry beneath. The flower buds are perhaps more striking than the flowers, being relatively large, almost globose and with many closely overlapping scale leaves. They have been aptly likened to drumsticks. Often associated with the pachystegias, mostly at inland sites, are two other Marlborough and North Canterbury endemics, *Brachyglottis monroi* and *Hebe hulkeana* with its freely branched inflorescences of small mauve flowers. Other wider ranging shrubs are the dwarf kowhai (*Sophora prostrata*), the leafless *Clematis* (*C. afoliata*), the densely twiggy *Corokia cotoneaster* and other small, drought-resistant shrubs. The latter group of species is also frequent on dry rocky outcrops further south in inland Canterbury and Otago.

## Sand Dunes[125]

Sandy beaches built up from the sea and backed by sand dunes that have been formed by wind are a feature of much of the coastline of both islands. These beaches, regularly disturbed by tides, rarely support any plants and even the dune ridge immediately behind the beach is a difficult habitat for them. The sand is unstable, frequently dry near the surface and with a high salt content from wind and spray. A few plants however, known as sand binders, colonise such dunes and often completely cover them with foliage. The sedge pingao (*Desmoschoenus spiralis*) (Fig. 88), now becoming rare, is the most robust and colourful of these with its coarse, narrow, curving leaves coloured yellow to orange-brown. It has widely spreading and branching subsurface rhizomes, which give rise to cord-like roots penetrating deep into the sandhill. The silvery grass *Spinifex sericea* is equally common, but in this case the more slender rhizomes spread and branch over the surface of the sand. The seed heads of *Spinifex* are distinctive in that they bear quite long, spherically arranged, flexible spines, which enable the heads to bowl along with the wind over the beaches and dunes for considerable distances. The introduced marram grass (*Ammophila arenaria*) is now also common as a sand binder.

As a dune becomes consolidated by the sand binders and particularly

*Figure 88* Robust leading shoots of pingao (*Desmoschoenus spiralis*) spreading through a sand dune. Photo: J. Miles.

when new sheltering dunes are formed seaward of it, other species less tolerant of instability enter in. These include the herbaceous *Tetragonia trigyna* (New Zealand spinach) and the *Convolvulus*-like *Calystegia soldanella*; the low shrubs *Pimelea arenaria* and *Coprosma acerosa,* which has interlaced mattress-like orangey stems bearing minute leaves.

Eventually a variety of taller shrubs, robust herbs and climbers common in open inland situations may form dense communities: toetoe (*Cortaderia fulvida*), *Phormium tenax,* species of leafless broom (*Carmichaelia*), matagouri (*Discaria toumatou*), species of *Cassinia,* manuka (*Leptospermum scoparium*), the introduced shrubby *Lupinus arboreus* and others. Given continuing stability and sufficient time a progression to forest can take place.

In the hollows between dunes, damp or even swampy conditions often prevail. The first plants to establish in such moist and relatively sheltered situations are such small herbs as *Gunnera arenaria, Ranunculus acaulis* and the sedges *Isolepis nodosus* and *Carex pumila*. The taller jointed rush (*Leptocarpus similis*) succeeds these, followed sometimes by manuka. The more extensive swampy interdune areas may eventually support swamp forest.

## Shingle Beaches and Fans[125]

Along some parts of the coastline, beaches are stony rather than sandy. They provide a more stable site for plants and as habitats they are similar to the sometimes adjacent shingle fans that result from the erosion of sea cliffs.

Nearest the sea are scattered succulent herbs such as *Apium prostratum* and *Senecio lautus*. Further back we find larger herbs like *Phormium cookianum,* the sedges *Isolepis nodosus* and *Cyperus ustulatus,* the white-flowered New Zealand species of true linen flax, *Linum monogynum,* and patches of the strange, leafless shrub, *Muehlenbeckia ephedroides*. On the terraces behind the beaches and on shingle fans there may be a close cover of densely twiggy, small-leaved shrubs belonging to several genera — *Coprosma propinqua, C. rigida, Melicytus crassifolius, Muehlenbeckia astonii* and hummocky shrubby forms of the lianes *Muehlenbeckia complexa* and *Rubus squarrosus*. In this situation *R. squarrosus* is quite leafless and beset with yellow prickles. In the Cook Strait area the grey-leaved spaniard (*Aciphylla squarrosa*) is also a common plant on fans.

## Salt Marshes and Meadows[125]

On sheltered low-lying coasts in muddy estuaries and lagoons, salt marshes frequently or occasionally flooded by tides, grade landwards into salt meadows rarely reached by the sea. Nearest the sea, a zone that is under water for part of each day during spring tides has a close cover of the glasswort, *Sarcocornia quinqueflora*. Further from the sea and at a slightly higher elevation reached only by the highest spring tides, the next zone is dominated by taller rush-like plants including the true rush *Juncus maritimus* and the jointed rush *Leptocarpus similis*. Often backing this zone is a fringe of the small-leaved twiggy shrub *Plagianthus divaricatus,* a strange relative of the much larger ribbonwood (*P. regius*) found in inland forests. Furthest from the sea the salt meadow is rarely reached by the tide, but owes its moderate saltiness principally to salt-laden gales. It has a turf of low-growing, mainly succulent herbs including *Apium prostratum, Selliera radicans* with its lobelia-like flowers, *Samolus repens* and *Leptinella (Cotula) dioica. Triglochin striatum* and *Lilaeopsis* spp. are also common here.

## Serpentine Vegetation

What is generally known as serpentine vegetation occurs in localised areas throughout the world on soils derived from ultramafic rocks. These soils provide a difficult medium for plants, as they are highly infertile with low levels of potassium, phosphorus and calcium and often high concentrations of iron and magnesium and the toxic metals nickel and chrome.

As a result, in forested areas serpentine vegetation often has a very different appearance from the vegetation adjacent to or even surrounding it (Fig. 89). The serpentine vegetation is more open with fewer, somewhat stunted trees and more shrubs and herbs. Some of the species on serpentine, although often reduced in stature, are shared with non-serpentine forests, shrublands and grasslands; others are unique and are known as serpentine endemics.

In the South Island there is a belt of ultramafic rock, known as the 'mineral belt', which runs through D'Urville Island, then to the east of Nelson and ends south of the Wairau River.[127] A similar belt is found far to the south at the northern fringes of Fiordland. The two belts are considered to have been continuous at one time until separated by lateral movement along the alpine fault.

In the North Island the only occurrences of ultramafic rock are at the northern tip along a few kilometres of the Surville Cliffs near North Cape and at the tip of the peninsula a little to the south.

The serpentine flora at the Surville Cliffs has been investigated recently and 144 species have been recorded.[128] Two unnamed shrub species of *Hebe* and *Coprosma* are considered to be endemic as well as 10 varieties, mostly of shrubs. There are patches of low forest on the cliffs comprising common local species of which pohutukawa (*Metrosideros excelsa*) is the dominant. Elsewhere the cover consists mainly of low shrubs draped with the string-like stems of the parasitic *Cassytha paniculata*. The only conspicuous herbs are a toetoe (*Cortaderia splendens*), *Astelia banksii* and *Phormium tenax*.

A curious feature of a number of the shrubs at this locality is their 'semi-lianoid' habit. Their 'elongated stems do not climb, but scramble or trail down the cliffs through other plants'. Most of the shrubs exhibiting this habit are considered to be varieties of wider ranging species.

## The Effects of European Settlement[129,130]

This book is largely concerned with the native flora of New Zealand which has been evolving here for many millions of years. Changes brought about by Europeans, in particular the establishment of introduced plants or 'adventives', has occupied just a moment of geological time. It is appropriate to consider these recent changes in this chapter as they tend to involve lowland open habitats.

A major aim of the European settlers was to establish farms, particularly sheep farms for the production of wool and meat. Naturally the grasslands of the eastern South Island and southern North Island were exploited first, since there was no forest to be cleared and less possibility of conflict with the Maori, whose numbers were lower in the South Island.

The initial development of sheep runs on the natural grasslands took place from the 1840s to the 1860s. At first the sheep were grazed on the native grasses, but as these were mostly coarse and not very palatable a practice developed of periodically setting fire to the grassland to induce soft new growth from the tussocks. This resulted in loss of humus and reduced the fertility of the soil, so eventually the degraded short tussock grasslands of the Canterbury Plains and parts of Otago

*Figure 89* View of the Olivine Range showing a contact between ultramafic rock (left) and schist (right). On the schist there is the normal forest to alpine vegetation sequence, on the ultramafic rock this is absent and there is a sparse cover of shrubs and herbs without a clear altitudinal zonation. Photo: P. F. Newsome.

were replaced with European pasture grasses or grain crops.

In the higher country burning continued, and this, combined with overgrazing and the depredations of other introduced animals, particularly the rabbit, led to degeneration of the native vegetation. In places, tall tussock grassland changed to short tussock grassland, and elsewhere short tussock grassland became virtual desert. The situation has improved somewhat in recent years with the cessation of burning and the control of rabbits.

After the end of the New Zealand Wars in the North Island in the 1870s, extensive forest clearance began. In the north, the main purpose was to obtain valuable construction timber from the forests containing kauri. Elsewhere, obtaining timber was secondary to land clearance for sheep or dairy farming, and the forests were often destroyed by fire.

The fernlands and shrublands of the eastern North Island were also cleared for farming, but this was not successful on the volcanic plateau until the cobalt deficiency of the soil was corrected. In this region extensive plantations of exotic pines have been and are still being established.

As the higher mountain areas are unsuitable for human settlement, it might be expected that the alpine vegetation there would have been little altered since the arrival of human beings. It has been less altered, certainly, but introduced wild animals, notably deer and in places goats and thar, have reduced palatable species and interfered with natural regeneration. As with the rabbits at lower elevations, some success has been achieved in controlling the numbers of these animals, particularly since the advent of deer hunting by helicopter. In places in the mountains some alpine plants which had become rare are now re-establishing.

Deer have also been a problem in the forests, as has the introduced Australian opossum, which may entirely defoliate some palatable species such as northern rata (*Metrosideros robusta*) and five-finger (*Pseudopanax arboreus*).

Since European settlement a considerable number of plants from other countries, most notably Europe, have been deliberately or accidentally introduced. Some of these have established themselves in the wild, mainly in the settled lowlands, and are known as the adventive flora. Many of the adventives are herbaceous, often annual weeds of disturbed and waste places, and are not individually conspicuous. Some cover more extènsive areas and may form pure associations. Examples are gorse, particularly about Wellington; blackberry and broom in many areas; lupin on sand dunes; california thistle and ragwort in cultivated areas; nassella tussock in North Canterbury; sweet briar and sheep's sorrel on the drier eastern South Island; thyme in Central Otago; and in the northern North Island water hyacinth, which is presenting a problem by blocking streams and lakes. The more cold-tolerant water weeds *Elodea* and *Lagarosiphon* appear to be spreading through both islands.

The adventives seemed so aggressive and successful that some early observers feared that the native flora was endangered. Armstrong[131] wrote: 'The rapidity with which foreign plants become naturalised in New Zealand is indeed a most surprising and extraordinary circumstance, as it must be quite evident to every observer that the introduction of these European plants will certainly result in the extermination of the indigenous flora, and that at no distant period of time.' Cockayne[132] disagreed with this view. He considered that few adventives were successful where native vegetation was well established. This is still generally the case. Examples of the few exceptions are heather, which is well established in Tongariro National Park, *Pinus contorta*, which is an increasing problem in the same area, and *Clematis vitalba*, at a number of places throughout.

CHAPTER 8

# Alpine Plants

Internationally, the treeline is generally taken to be the lower limit of alpine vegetation. In New Zealand, where beech species are present, the treeline may be very clearly defined with the trees abutting directly onto either alpine tussock grassland, with or without an admixture of shrubs, or a narrow belt of shrubs which widens on steep, sunny, stony slopes and spurs. Where beech species are not present the treeline formed by montane conifer broadleaf forest species is both more diffuse and at lower altitudes than that of beech at comparable latitudes. Above such treelines there is usually a very wide zone of shrubs extending up slope for 100-300 m.[133] Some consider that this zone is climatically suitable for beech forest and so refer to it as 'subalpine shrubland' implying that the climatic or 'regional' treeline lies at the upper limit of the shrubs. As a modification of this view, Wardle[95] suggests that the regional tree line would lie at the upper limit of shrubs capable of becoming small trees. This implies that below such a line the shrubland is subalpine and above it alpine (Fig. 90).

It seems more convenient to treat the higher altitude shrublands as one entity and for this reason I use the non-committal term 'mountain shrublands'.

## Mountain Shrublands

Where the mountain shrubland zone is dense and extensive (Fig. 91) it is easy to be impressed both by the variety of species present and by their range in leaf form (from large and leathery to needle- or scale-like) and in colour (from dark green to yellow-green or even orange-red). Unfortunately, if no tracks are present, this pleasing first impression may soon be dissipated by the discomforts and frustrations of trying to make progress. The branches of some of the larger shrubs lie on and branch over the ground for considerable distances, so one alternates between trying to push through and between them and attempting to

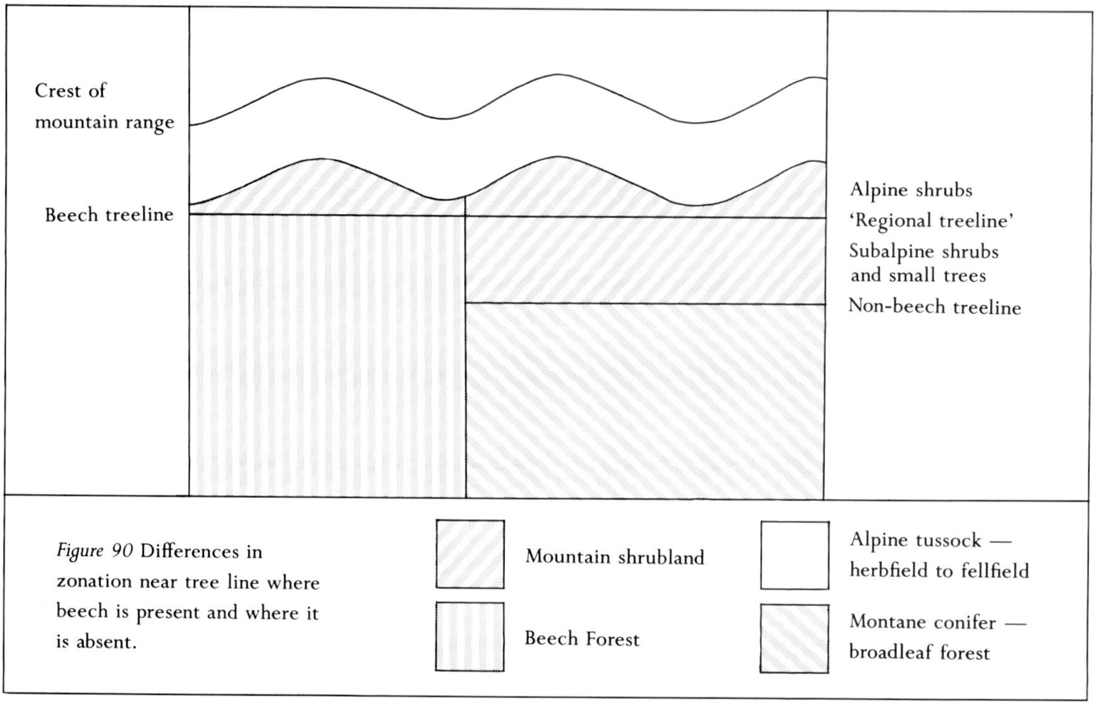


*Figure 90* Differences in zonation near tree line where beech is present and where it is absent.

*Figure 91* Mountain shrubland on Mt. Taranaki (Egmont). Beech (*Nothofagus*) forest does not occur on this mountain. Photo: J. W. Dawson.

scramble over their tops. After perhaps hours of snail-like progress it is a great relief to reach the open alpine vegetation above or the forest below.

### Subalpine Shrubs or Small Trees

In the lower or subalpine parts of the shrub zone the species composition depends on soils and aspect. In the Westland beech gap Wardle[134] recognises a number of subalpine low forest or shrub communities which are in part typical for New Zealand as a whole. They include:

#### *Hoheria glabrata* low forest

*Hoheria glabrata* or mountain ribbonwood colonises young, deep, moist, well-drained and often stony soils such as those provided by slips, talus slopes and alluvial fans. This small tree, being one of the few in New Zealand that is strongly deciduous, has attractive yellow to red leaves in the autumn. Its clusters of white flowers, each up to 4 cm across, are also a striking feature. *Olearia ilicifolia* (the holly-leaved daisy tree) also contributes to the canopy and the ground beneath has a dense cover of the fern *Polystichum vestitum*.

#### *Dracophyllum-Olearia* Low Forest and Scrub

This occupies similar but older sites than *Hoheria glabrata* low forest. Dominance is shared by the mountain ribbonwood (low forest only), *Dracophyllum longifolium* (scrub only), *D. traversii* and *Olearia lacunosa,* the last two being up to 7 m tall. The genus *Dracophyllum,* although belonging to the dicotyledon class of the flowering plants has long, narrow, parallel-veined leaves similar, in the larger examples, to those of such monocotyledons as the cabbage tree (*Cordyline australis*) or even the pineapple. Indeed, mountain groves of *Dracophyllum traversii* are sometimes known locally as 'pineapple forests'. This species with its candelabra-like form, and its sword-like leaves forming a deep litter on the forest floor, is certainly the most distinctive member of the community.

At higher altitudes in the south-west of the South Island *Dracophyllum traversii* is replaced by the smaller *D. menziesii* and the equally distinctive *D. fiordense,* which has an often unbranched trunk up to 2 m high topped by an almost spherical mop of strongly downwardly curved leaves (Fig. 92). *Dracophyllum longifolium* is one of many shrub species with smaller needle leaves and a heath-like appearance. Indeed the family

*Figure 92* A group of unbranched shrubs of *Dracophyllum fiordense* together with needle-leaved Dracophyllums and other shrubs. The grass is *Chionochloa flavescens*. Fiordland. Photo: Jane Maxwell.

to which *Dracophyllum* belongs, Epacridaceae, is closely related to the heath family, Ericaceae.

*Olearia lacunosa* also has narrow, although net-veined leaves, which are somewhat like those of juvenile lancewood (*Pseudopanax crassifolius*). Some of the other shrub species with relatively large thick leaves are *Griselinia littoralis* (broadleaf), *Pseudopanax colensoi* (mountain five-finger) and *Olearia colensoi* (leatherwood). *Pseudopanax simplex* has somewhat smaller, thinner leaves and *Myrsine divaricata* and *Coprosma pseudocuneata* have very small leaves. Prominent ground plants are *Blechnum 'procerum'* and the tussock-like *Astelia nervosa.*

Dominance of the shrub species varies from place to place in this community type, but it is at its most impenetrable where *Olearia colensoi* with its stiff almost cardboard-like leaves predominates, as in the North Island ranges, the westernmost coastal ranges in the South Island and in Stewart Island.

### Subalpine Heath-Scrub[135]

This occurs on old, leached infertile soils and is a low shrubbery from 0.5 to 1.5 m tall in which the small conifers *Halocarpus biformis* and *Phyllocladus aspleniifolius* var. *alpinus* (mountain celery pine) share dominance with *Olearia colensoi* and *Dracophyllum longifolium.*

Both the conifers also occur in the lowlands of Westland where they become small trees. In the greyish-green mountain celery pine, what appear to be fan-like or rhomboidal leaves are in fact flattened branchlets termed phylloclades.

These are some of the community types of subalpine shrubs and small trees, but there are a number of others related to special habitats or resulting from disturbances, whether caused by nature or by humans. Many of the species of these communities may grow in the montane forests especially towards their upper limits.

## Alpine shrubs

Above the actual or inferred treeline come shrubs that can be considered truly alpine, although many descend to lower altitudes in open habitats. Immediately above treeline on moderate slopes with a reasonably well developed soil such shrubs generally intermingle with large tussock grasses and other herbs in what has been termed tussock-shrubland. As altitude increases, the shrubs decrease and the herbs increase. On steeper slopes, such as spurs and ridges, with shallow, stony soils, species of shrubs mostly different from those of the tussock-shrubland may form a more continuous cover, but there is usually a herb component. On these rocky sites certain shrubs may extend to quite high altitudes, becoming shorter and more scattered as they do so.

It is in this higher shrubland zone that New Zealand's largest genus *Hebe* attains its greatest prominence. Of the approximately 60 commoner species of alpine shrubs 24 or 40 per cent are hebes, many of them exhibiting the symmetry of form and attractiveness of foliage — grey-green, yellow-green and shades of yellow and orange — that have made the genus so well known horticulturally in many parts of the world. The alpine hebes can be divided into two approximately equal groups contrasting strongly in growth habit. In the first group the leaves, although short and fairly broad, are spreading and often very precisely arranged in four vertical rows. The shapes of such shrubs may be so symmetrical and their surfaces so dense, that they appear to have been trimmed to shape. Appropriately, one of them has been named *Hebe topiaria*. The relationship of the hebes in this group with the willow-leaved species (koromikos) of the lowlands is quite evident, but with the second group of so-called 'whipcord' hebes this is not so; they resemble instead the quite unrelated scale-leaved conifers (Fig. 93). It is no surprise to learn that one of them is named *Hebe cupressoides*. Both growth forms are represented in tussock-shrubland and on stony ridges.

Similar in general appearance to the hebes with spreading four-ranked leaves are a number of species of *Pimelea* or New Zealand 'daphnes'. The flowers with their pairs of stamens are also quite Hebe-like, but as a general rule the two genera can be distinguished by the presence (*Pimelea*) or absence (*Hebe*) of leaf and flower hairs. Pimelea leaves often feel quite silky to touch. Pimeleas occur both in grassland and on rocks.

Most of the alpine dracophyllums have needle leaves. Taller species in tussock-shrubland have a switch-like habit with narrow branching angles, others in similar habitats or on rocky ridges are more or less prostrate. During the colder period of the year especially, the dracophyllums add colour to the alpine scene when their leaves develop bright reddish brown hues.

A few gaultherias belonging to the Ericaceae or heath family also occur in alpine communities. *Gaultheria crassa* is a shrub up to 1 m high growing in a variety of sites. The clusters of small white flowers are bell-like and not unlike those of the dracophyllums. The leaves, however, are short, broad and quite thick. *Gaultheria depressa* var. *novae-zelandiae* is a prostrate shrub in open places remarkable for its relatively large white, pink or red, berry-like fruits often referred to as 'snow berries'. The

fleshy parts of the fruits are actually the persistent and enlarged sepals of the flowers (Fig. 94). *Pernettya macrostigma,* which belongs to a genus closely related to *Gaultheria,* is a scrambling, interlaced shrub of poorly drained areas in tussock grassland. It has an even more remarkable fruit than that of *Gaultheria depressa,* as both the persistent sepals and the seed capsule within them are fleshy. Natural hybrids between *Pernettya macrostigma* and a *Gaultheria* have bright red fruits in which the plump sepals form a crown-like cup around the capsule.

*Figure 93 (left) Hebe tetragona.* A 'whipcord' Hebe.
Photo: M. D. King.

*Figure 94 (right) Gaultheria depressa.* The upper fruit has had part of the fleshy calyx removed to reveal the seed capsule.
Photo: M. D. King.

All except one of the relatively few shrubby alpine coprosmas have very small leaves and slender twigs. Three of these are prostrate and scrambling, *C. cheesemanii* and *C. crenulata* in damp places in tussock grassland and *C. depressa* on rocky sites. *Coprosma pseudocuneata* with its attractive, strongly recurved, yellowish green leaves has an erect habit and extends to quite high altitudes in sheltered rocky sites. The one large-leaved *Coprosma, C. serrulata,* grows in shaded tussock shrubland

and also on shaded rocky bluffs. Its leaves are thick and leathery with prominent veins.

The Compositae or daisy family is represented by a few species belonging to three genera.

*Brachyglottis (Senecio) bidwillii* is widespread in tussock shrubland and on rocky bluffs. Its leaves are rounded and very thick with a thick, felt-like mass of hairs on their undersides. Two other species are more localised, *B. adamsii* in adjacent parts of the North and South Islands, and *B. revolutus* in the south-west of the South Island.

The genus *Helichrysum,* which includes the flowers known as 'everlastings', has several shrubby alpine species in the drier eastern mountains of the South Island where they grow on rocky outcrops and ridges. They all have a similar distinctive appearance with scale-like, but somewhat swollen, strongly convex leaves closely pressed to the stems. The rounded backs of the leaves may be very shiny and each leaf is surrounded by a dense mass of white hairs. The most robust and striking species is *Helichrysum coralloides* with cylindrical stems up to 1 cm in diameter.

*Cassinia vauvilliersii* may be common in tussock-shrubland throughout New Zealand as well as in open vegetation at lower altitudes. It is an erect shrub, often with a distinct yellowish colouration from the hairs on the undersides of the small leaves.

*Melicytus (Hymenanthera) alpinus* of the violet family is a small shrub with short, stiff, interlacing twigs which are almost spiny at the tips. The species occurs throughout the South Island and is most frequent on rocky ridges and outcrops.

*Myrsine nummularia* is a small-leaved, prostrate, thin-stemmed shrub found throughout the New Zealand mountains where it favours rock outcrops and open places in tussock grassland. Its most striking feature is the bright violet-blue colouration of the berries.

Three dwarf conifers may be frequent throughout tussock shrubland. Snow totara (*Podocarpus nivalis*) forms low bushes in this community, but is also common at scree margins and on moraines at lower elevations. The pigmy pine (*Lepidothamnus laxifolius*), sometimes called the smallest conifer in the world, forms even lower mat-like patches in poorly drained sites and the mountain celery pine (*Phyllocladus aspleniifolius* var. *alpinus*) is still erect in habit though shorter than it is in the subalpine shrublands.

## Tussock Herbfield[136]

This type of alpine vegetation occupies a zone about 500 m wide above the treeline where soils of adequate depth are present. The term 'herbfield' as applied to northern hemisphere mountains is defined as a complete cover of herbaceous alpine plants, but in New Zealand the

*Figure 95* Red tussock grassland on Mt. Taranaki (Egmont). Tussocks with seed heads in foreground.
Photo: J. W. Dawson.

modified term tussock herbfield is preferred. This is because the dominant and conspicuous large tussock grasses belonging to the genus *Chionochloa* (Figs. 95, 96), although they lack woody tissues, are often as tall as or taller than some of the alpine shrubs.

**Snowgrasses**

The larger alpine tussock grasses, or 'snow grasses' as they are called, are frequently a metre high and sometimes as wide. Under good conditions they may reach head height at up to 2 m. Growth is slow and some of the larger specimens are estimated to be several centuries old.[137] The more or less hemispherical form of many of the snow tussocks confers on the mountain slopes a very distinctive and attractive texture reminiscent of that of cirrus clouds or, when the tussocks are tossed by the wind, waves of the sea. The colour of a snow tussock landscape is generally not green, but, depending on the species, ranges from a pale straw colour through shades of brown to a distinctly reddish shade. The lack of greenness results partly from the regular dying back

*Figure 96* The snow tussock (*Chionochloa macra*) with flower heads of the spaniard (*Aciphylla scott-thomsonii*). Old Man Range, Otago.
Photo: J. W. Dawson.

of the leaves from the tips and partly to the presence of masking pigments. Heavy flowering seasons of the snow grasses are sporadic and often coincide with those of the beech species. This suggests that snow grasses too require a warm summer preceding flowering for the initiation of flower buds.

The flower and seed heads of some of the larger snow tussocks are quite diffuse, and their very slender stems and small, scattered, pale, flower or seed heads give an insubstantial misty effect similar to that of the garden *Gypsophila,* so popular in flower arrangements. On the wetter western mountains of the South Island the broadleaved snow tussock (*C. flavescens*), can grow sometimes 2 m high, and dominates in the zone 200 m above treeline. The mid-ribbed snow tussock (*C. pallens*) is often present, generally on younger, better drained soils. Both species are also common on the North Island axial ranges. With increasing altitude on the wet South Island mountains the two large snow tussocks gradually give way to the much smaller curled snow tussock (*C. crassiuscula*).

On Mt. Taranaki (Egmont) and the central volcanoes in the North

*Figure 97* Dense cover of red tussock (*Chionochloa rubra*) on a swampy valley floor. Near Boulder Lake, north-west Nelson. Photo: J. W. Dawson.

Island the only snow tussock present is red tussock (*C. rubra*), which occupies a wide range of habitats in the herbfield zone. In the South Island on the wetter mountains red tussock is abundant only at or below the treeline on poorly drained and frosty flat valley floors. With its large size and strong reddish colouration it provides an impressive sight (Fig. 97).

On the drier eastern South Island mountains in the southern half, the dominant snow tussock in the low alpine zone is the narrow-leaved snow tussock (*C. rigida*). This gives way at higher altitudes to the smaller recently described slim snow tussock (*C. macra*). Only slim snow tussock extends north of the central South Island and it then occupies a broader altitudinal zone.

Among the smaller species of *Chionochloa,* which are found mostly at higher levels in herbfield, carpet grass (*C. australis*) is worth a special mention. As its common name indicates it does not have a tussock habit, but forms thick, often extensive swards. The needle-like leaves are dark green and shiny and as they are also slippery and tend to lie downhill, they need to be walked on with care. Carpet grass is found on wetter mountains in the northern third of the South Island.

Where snow tussocks are tall and dense near the treeline one might think there would be little room left for other alpine herbs; in fact there are quite a number. Some are small and inconspicuous, enjoying the shelter and tolerating the shade of the tussocks. Others are much more conspicuous and could be termed large or even giant herbs, approaching or exceeding the tussocks in height.

## Spaniards

Notable among the large alpine herbs are the spaniards or speargrasses belonging to the genus *Aciphylla* of the Umbelliferae (carrot family).[138] Nothing could look more unlike a carrot plant than the larger spaniards of tall tussock grassland. In their tussock-like clumps the large leaves are deeply divided into hard, rigid segments tipped by needle-sharp spines. The flower, and later seed, heads are equally unusual in that, instead of being broad and open as is more typical for the family, they are dense, narrow and lance-like. The individual flower clusters, densely aggregated on the upper parts of the lances, are exceeded in length by their associated bracts. These bracts have segments as spiny as those of the leaves. Having suffered while collecting specimens of such spaniards for study, I am inclined to agree with the suggestion that the excessive spininess is a defence against browsing animals (and botanists), which in pre-human times in New Zealand would have been the moas. However, despite their unpleasant characteristics, many of the spaniards are striking in appearance and, when strongly coloured, decidedly handsome as well. Some species have grey-green leaves and flower heads ranging from green to pale yellow, others have both leaves and flower heads with a quite strong yellow to orange colouration as, for example, in the golden spaniard (*Aciphylla aurea*) (Fig. 98).

Most of the large species of spaniard grow in moist tussock herbfield or open mountain shrubland. In the very wet south west of the South Island is the appropriately named *Aciphylla horrida*. Further east, but still in moist situations, the grey-green *A. scott-thomsonii* is prominent (Fig. 96). This species has the distinction of being the largest, with seed heads

*Figure 98* The golden spaniard (*Aciphylla aurea*) scattered through tussock grassland comprised of narrow-leaved snow tussock (*Chionochloa rigida*). Mt. St. Bathans, Otago. Photo: J. W. Dawson.

sometimes as much as 4 m high. Other large species occur further north with a few extending through the North Island ranges. One of the latter, *A. colensoi,* is notable for the bright yellow to orange midribs of its leaf segments.

The golden spaniard, *A. aurea,* is often prominent in drier habitats in the eastern South Island and is characteristic of narrow-leaved snow tussock grassland (Fig. 98). In higher herbfield where the plant cover is shorter the larger species of *Aciphylla* give way to smaller forms. Some of the smaller forms are spiny with narrow inflorescences; others are more like the related *Anisotome* with soft leaves and broad inflorescences.

It is difficult to understand why narrow dense inflorescences should have evolved in *Aciphylla*. It could be seen as a protective device as the short flower clusters are readily shielded by the spinescent bracts, but similar (although not spiny) inflorescences occur in quite unrelated plants elsewhere in the world — the tree senecios and lobelias of the central African Mountains, the *Puya* (Bromeliaceae) of the northern Andes, *Echium* in the Canary Islands and the grass trees (*Xanthorrhea*) in Australia, to name a few.

### Mountain Buttercups[139]

The largest and best known of these is *Ranunculus lyallii,* the so called 'Mount Cook Lily'. It seems that any large herb, especially if it has large, white flowers, as in this case, is inevitably and often erroneously

*Figure 99* Flowers of *Ranunculus lyallii* at Arthurs Pass, Canterbury.
Photo: J. W. Dawson.

*Figure 100* Group of alpine plants at Arthurs Pass, Canterbury. The large circular leaves are *Ranunculus lyallii*. Two mountain daisies are present, *C. semicordata* below and *C. armstrongii* above. The herb with very divided leaves is *Anisotome haastii* and the long silvery leaves at the top belong to *Astelia nervosa*. The needle-leaved shrubs are *Dracophyllum longifolium*.
Photo: J. W. Dawson.

termed a 'lily'. *Ranunculus lyallii* is most admired for its large, pure-white flowers up to 6 cm across (Fig. 99). It is equally remarkable for its large, rather floppy, almost circular leaves up to 30 cm in diameter (Fig. 100). Plants of this species may reach 1 m in height among tussocks or shrubs, particularly in moist places, throughout the South Island mountains, except in the far north, and in Stewart Island. Other smaller, but still large, species have bright buttercup yellow flowers, for example the variable *Ranunculus insignis* throughout the axial ranges of the North Island and in the northern half of the South Island and *R. nivicola* in the tussock herbfield of Mt. Taranaki (Egmont) and also on the central North Island mountains. *Ranunculus verticillatus,* which is also yellow-flowered, is more slender and less conspicuous, but nevertheless common in the southern North Island and northern South Island. It often grows through, and is supported by, shrubs and snow tussocks.

### Ourisia

The ourisias are sometimes called mountain foxgloves. This is a common name with at least some validity since foxgloves and *Ourisia* belong to the same family, the Scrophulariaceae.

The two large species of *Ourisia* found in tussock herbfield, like the large *Ranunculus* species, have leathery or somewhat fleshy, dark green leaves. The flowers are white with yellow centres, 2–3 cm across and have the five petals fused at the base. The flowers are distinctively arranged in spreading circles or whorls, one above the other. *Ourisia macrophylla* with two subspecies is found on wetter mountains in the North Island and *O. macrocarpa* also with two subspecies in similar sites throughout the South Island.

### Gentians

To those familiar with the brilliantly blue, deeply bell-shaped gentians of the northern hemisphere, the New Zealand versions must come as a surprise. Their flowers are like sprinklings of snow among the tussocks and in each flower the petals are much more deeply divided than those of northern species although still fused near the base (Fig. 101). The leaves are generally somewhat fleshy and sometimes both leaves and stems have a strong red or purple colouration masking the green. The flowers are usually pure white, but sometimes have purple veins. The distinctions between many of the New Zealand species are still not clear, but several are prominent in tussock herbfield and shrubland. *Gentiana corymbifera* can be up to half a metre tall and is mostly found in drier eastern tussock grassland throughout the South Island mountains. The similar *G. montana* is widespread in the wetter, mostly western mountains of the South Island and in Stewart Island. *G. bellidifolia,* a somewhat smaller species, ranges throughout the mountains of both North and South Islands.

## Mountain Daisies[140]

Mountain daisies or celmisias are probably the most frequently encountered of our alpine herbs (Fig. 102). In other genera, two or three species may occur together at the same locality. Celmisias may have as many as five or six species occurring together. As befits the family Compositae, *Celmisia* flowers[141] are daisy-like and white with yellow to orange centres. In some species they may measure up to 7 or 8 or even 10 cm in diameter, The leaves are more diverse and remarkable than

*Figure 101 (below) Gentiana bellidifolia*. Mt. Peel, N. W. Nelson. Photo: J. W. Dawson.

*Figure 102 (right)* A group of celmisias in flower — *C. semicordata*. There are a few non-flowering rosettes of *C. traversii* also. Mt. Arthur, N. W. Nelson. Photo: J. W. Dawson.

the flowers. In some species they are thin and soft in texture and sometimes sticky, but in many others they are very firm with hard upper surfaces and in colour range from grey-green through yellow-green to dark green. The most notable and attractive feature of many *Celmisia* leaves is a dense and often thick felt of hairs on the undersides and sometimes the upper sides as well. This tomentum, as it is termed, may be pure white, cream to buff or even bright rusty brown.

Some celmisias of tussock herbfield have their leaves in one or more rosettes close to the ground. Others, which form in patches, have closely branched, spreading stems with more dispersed leaves. The stems of some of the latter can be fairly woody, and when they are semi-erect, the plants could be described as dwarf shrubs.

*Celmisia semicordata* (Fig. 102) has the largest leaves and flowers of the genus. The leaves are in large rosettes and are silvery green with white undersides. The species is common in both open shrubland and tussock herbfield on wetter mountains throughout the South Island except the far north. *Celmisia verbascifolia,* with a similar range to *C. semicordata,* and *C. traversii* in the north-west and south-west of the South Island are similar in habit, but somewhat smaller than *C. semicordata. Celmisia verbascifolia* is notable for the bright purple petioles and midribs of its leaves and *C. traversii* for the deep, velvety rusty red tomentum on the leaf undersides. The commonest *Celmisia* in New Zealand is *C. spectabilis,* which in some of its forms is like a small *C. semicordata.* It occurs throughout the mountains of the North and the northern half of the South Island in both wet and dry tussock grassland. It becomes particularly abundant following fires.

Three species are similar in having tufts of long, narrow, pointed sword-like leaves, which leads to them sometimes being mistaken for spaniards (*Aciphylla*). *Celmisia lyallii* ranges along the eastern sides of the South Island mountains in *Chionochloa macra* grassland, and *C. petriei* and *C. armstrongii* together span the wetter western mountains of the South Island, *C. petriei* in the north-west and south-west and *C. armstrongii* in between.

Of the many smaller leaved species which form spreading mats a metre or more in diameter, *Celmisia incana* is probably the most striking as it often has a completely white tomentum of hairs on both sides of the leaves as well as the flower stems. It ranges throughout the North Island mountains and continues to the middle of the South Island.

Three genera of monocotyledons other than grasses may also be prominent in tussock herbfield and shrubland.

### Bulbinella

Species of this genus may be abundant on moist, shady slopes in tussock herbfield, their heads of starry yellow flowers often providing something unusual in New Zealand mountain landscapes — a mass of solid colour. Their leaves are tufted and long and narrow. *Bulbinella hookeri* ranges from the North Island mountains to north Canterbury; *B. angustifolia,* sometimes inappropriately known as 'Maori onion', ranges through the drier eastern mountains of the South Island from North Canterbury southwards; and *B. gibbsii* in the wetter mountains of the southern North Island, southern South Island and Stewart Island.

### Mountain Flax

The term 'flax' for the genus *Phormium,* although firmly established, is yet another inappropriate common name as *Phormium* and the true linen flax (*Linum*) have in common only the possession of useful fibres. Mountain flax is *Phormium cookianum* and its tufts of narrow, leathery leaves may be conspicuous in poorly drained places in tussock herbfield and shrubland. The flowers in its quite tall heads vary in colour, from plant to plant, from yellow to dark reddish purple.

### Astelia

The thirteen New Zealand species of *Astelia* are about equally divided between the forests and the mountains. The flower heads of the smaller alpine species are short and inconspicuous, but their bright red to orange berries are strikingly attractive. Two large species may be conspicuous in wet places in tussock herbfield and shrubland. *Astelia nervosa* (Fig. 103) found throughout the country, has tufts of leaves up to a metre or more in length, varying from pale green to silvery white. In the latter case the leaves look as if covered with frost. *Astelia petriei,* with shorter, broader, pale green leaves, grows on the higher rainfall mountains of the western South Island. The smaller, patch-forming *Astelia nivicola* has a similar range but at higher altitudes in herbfield. The smallest herbfield species is the grasslike *Astelia graminea,* which is especially common among carpet grass (*Chionochloa australis*) in the northern South Island.

## Cushion Bogs[142]

Bogs develop in the subalpine and low alpine zones where drainage is poor.[143] Suitable sites are hollows formed by glaciers and flat areas under high rainfall conditions on glacial terraces, mountain passes and flat topped mountain ridges. Where rainfall is high, rain may provide most of the water in the bogs, but in drier eastern areas drainage from the surrounding slopes makes the major contribution. When the water table is near the surface, the dominant plants in the bogs are of cushion form. Cushion plants are freely but closely branched; and the ultimate twigs, with their upper living and lower dead leaves, are so closely pressed together lengthwise that the exposed tips of the living leaves form a firm, continuous, often unyielding surface. Many cushion plants can be stood on without suffering any visible effects. The leaf tips of each ultimate branchlet may form a circular pattern or be compressed into an hexagonal shape. Depending on the species the cushions range from a few centimetres to a metre or more in diameter. They are broadly convex to almost hemispherical in form and are separated by shallow to quite deep hollows.

A number of sedges and related plants occur in mountain bogs throughout the country. The *Centrolepis* and *Gaimardia* species form soft moss-like cushions; *Oreobolus pectinatus* with its leaves distinctively arranged in two rows in one plane forms a firmer cushion while other species of the genus form flat mats. Of other cushion plants, *Donatia novae-zelandiae* forms much larger, broadly convex, extremely dense cushions whose dark green leaf tips contrast strongly with the numerous, small white flowers in season (Fig. 104). It occurs from the southern North Island southwards and also in the Tasmanian mountains.

As well as the cushion plants there may also be prostrate but less compact dwarf shrubs including *Dracophyllum muscoides, D. politum* and the pigmy pine (*Lepidothamnus laxifolius*). In the hollows between the cushions and prostrate shrubs are a variety of small sedges and mosses and small species of several large genera including *Astelia linearis* with its red jelly bean-like fruits, *Celmisia glandulosa, Gentiana lineata* (in the far south) and dense mats of *Coprosma perpusilla* (formerly *C. pumila*) Several species of the insect catching sundews (*Drosera*) with their glistening leaf glands are also frequently present.

In the highest parts and margins of bogs, red tussock (*Chionochloa*

*Figure 103 (opposite above)* *Astelia nervosa* in snow tussock (*Chionochloa pallens*) grassland. Mt. Holdsworth, southern North Island. Photo: J. W. Dawson.

*Figure 104 (opposite below)* A dense cushion of *Donatia novae-zelandiae* with numerous flowers. Garvie Range, Southland. Photo: J. W. Dawson.

*rubra*) may often be found. In the lowest, wettest parts, soft masses of the yellow-green bog moss (*Sphagnum*) predominate.

## Fellfield

On the higher, steeper mountains, particularly of the South Island, herbfield gives way with increasing altitude to fellfield (Fig. 105). As a result of the severe environmental conditions in this zone, plants of this type of vegetation are both sparse and specialised. They are subjected to low average temperatures, heavy frosts, deep snow for part of the year, violent winds, and at times strong sunshine, which may raise the temperature of the extensive areas of exposed rock very considerably. Erosion by frost action and wind is quite rapid particularly on the greywacke sandstone of the Southern Alps and other axial ranges. The resulting angular fragments may form a thin layer of debris on ridge crests and sides; but gravity, assisted by wind and, more dramatically, avalanches, constantly moves the products of erosion downslope, so there is little opportunity for the development of even thin soils. Rock outcrops and bluffs in this zone provide more sheltered and secure sites for plants. Nevertheless a number of them do establish away from outcrops, especially in deeper debris on moderate to gentle slopes. Of these, some are small prostrate *Hebe* shrubs which are also to be found on rock outcrops. They are not of the whipcord type, but have small rounded leaves arranged in attractive patterns. *Hebe haastii* and *H. epacridea* (Fig. 106) are yellow or orangey-green; *H. petriei* is grey-green. Small herbs are more common — several diminutive grasses and sedges, and one to several species of *Epilobium, Ranunculus, Celmisia, Gentiana, Myosotis* and *Parahebe*. Small cushion plants also occur; these include *Hectorella caespitosa,* the softly pubescent species of *Chionohebe, Phyllachne colensoi*, and sometimes extensive mats of the silver grey *Celmisia sessiliflora*.

Rooted into clefts of rock outcrops or other stable sites are a number of rather larger and more striking plants. Hebes are prominent and include both whipcord and non-whipcord species. Notable among larger herbs are the grey-green, hairy *Ranunculus buchananii* with large, white flowers and the yellow-green *R. grahamii* with large yellow flowers. The former grows in the south-west of the South island, the latter only in the vicinity of Mt. Cook.[144] Attractive as well as curious, because of their similarity to the edelweiss of the Swiss Alps, are two species of

*Figure 105 (above)* Fellfield on the Black Birch Range, Marlborough.
Photo: J. W. Dawson.

*Figure 106 (left) Hebe epacridea.*
Photo: M. D. King.

*Leucogenes* — *L. leontopodium* throughout the North Island and the north-eastern South Island and the smaller leaved *L. grandiceps* throughout the South Island and Stewart Island. The flowers of both species are white and woolly with yellow to orange centres.

Most outstanding in these situations, however, are a number of large cushion plants, some of which, because of their hummocky form and woolly pubescence, are known as 'vegetable sheep'. These larger cushion plants are extremely firm in texture and sometimes a metre or more in diameter. The vegetable sheep belong to the daisy family (Compositae) and they are essentially shrubs — as is evidenced by the tortuous woody stems revealed by the erosion of dead plants. The living leaves of branch tips at the cushion surface closely invest one another as permanent buds, while the dead interior leaves break down into a blotting paper-like, water retentive mass, which fills the spaces between the branches. The Canterbury vegetable sheep (*Raoulia eximia*), found on drier eastern mountains from mid-Canterbury to north Otago, is

*Figure 107* A rock in fellfield on the crest of the St. Arnaud Range, northern South Island. Frost action has formed fissures in the rock where small cushions of *Colobanthus canaliculatus* (left) and *Raoulia bryoides* (centre and right) have established.
Photo: J. W. Dawson.

*Figure 108* Large cushion of the Marlborough vegetable sheep, *Haastia pulvinaris*. Mt. Cupola, Nelson Lakes National Park. Photo: J. W. Dawson.

*Figure 109* Close view of a portion of a *Haastia pulvinaris* cushion showing the branchlet tips closely invested by woolly leaves. Photo: J. W. Dawson.

grey-green in colour and the velvety leaf buds are often compressed into more or less hexagonal shapes. Several other species of *Raoulia* form smaller cushions in fellfield (Fig. 107). The more handsome (in my view) Marlborough vegetable sheep (*Haastia pulvinaris*) is restricted to the drier mountains of the north east of the South Island. The large cushions are buff to pale yellow in colour and the leaf buds are larger and woollier than those of the raoulias (Figs. 108, 109).

Two spaniards (*Aciphylla*) are notable cushion plants in fellfield. *A. dobsonii* in South Canterbury and North Otago may form perfectly hemispherical cushions more than half a metre in diameter (Fig. 110), while *A. simplex* of Central Otago forms somewhat smaller cushions. Both species have thick, rigid leaves coloured bright orange-yellow.

The two flowering plants of fellfield which hold the altitude record are *Hebe haastii* and *Parahebe birleyii*, both having been recorded at 2900 m in the Mt. Cook region.[144] At 2800 m *Ranunculus grahamii* is not far behind. Above these altitudes only lichens and mosses cling to steep rocky faces in the zone of permanent snow.

*Figure 110* Hemispherical cushion of the cushion spaniard, *Aciphylla dobsonii*. Mt. St. Bathans, Otago. Photo: J. W. Dawson.

## Scree Plants

Screes or shingle slips might be regarded as a very special type of fellfield. They are widespread on the drier eastern mountains of Canterbury and Marlborough and some mountain peaks may have their mid and lower slopes completely buried in deep aprons of angular stones through several thousand metres of altitude (Fig. 111). These vast accumulations result from the already mentioned rapid disintegration of greywacke under high alpine conditions, and the relatively slow downward movement of the rock fragments in the absence of heavy rainfall.

The screes are poised at the angle of rest and are very difficult to traverse as the weight of a person will tip the balance and set the stones in motion. Altogether the screes seem an improbable habitat for plants.

Apart from their instability, the stones appear soil-less and arid and when the sun shines they can become very hot, raising the temperature of the air directly above very considerably. In fact when one manages to get out onto a scree it seems at first to be entirely plantless, then one small plant is spotted, then another and another. The plants are often grey like the stones, fleshy, and tend to be scattered fairly evenly across the scree several metres apart. At first they all appear the same, but closer inspection reveals several different and mostly quite unrelated species. About a dozen of these are found only on screes and there only away from the margins, in the deeper and most mobile parts. It is clear then that conditions are not as impossible as they seem. Beneath the surface the stones diminish in size quite rapidly to a sandy soil at a depth of about 40–50 cm. The extensive root systems of the plants, arising directly from the bases of leaf rosettes or from spreading rhizomes, branch widely through this soil, which is constantly wet from water seeping downslope. Thus scree plants do not in fact have a water problem, even when high temperatures at the scree surface lead to excessive water loss from the leaves. So far, so good, but what about

*Figure 111* A steep scree on the Craigieburn Range, Canterbury. The forest is of mountain beech (*Nothofagus solandri* var. *cliffortioides*). Photo: J. W. Dawson.

the instability problem? If a plant is lucky the scree will not move in its vicinity during the growing season, but if it does the scree plants are adapted to survive the event. If the layer of stones above the soil moves substantially, although the leaves and flowers or flower heads of a scree plant will be either buried or sheared off, its root system and rootstock or rhizome will survive intact and will be able to replace the lost parts. In some scree species the leaf petioles and flower stalks taper downwards so that their connections with the rootstock are almost threadlike and can break fairly readily. If these connections were thick and tough then the whole plant might be uprooted and die. The notable example of this type is *Lignocarpa carnosula,* a grey-green, fleshy plant with much divided leaves, which belongs to the carrot family (Umbelliferae). Its dry seed heads break off at the weak connection and blow around on the scree surface like a tumbleweed, scattering seeds as they go.

*Figure 112* The penwiper plant, *Notothlaspi rosulatum*. Porter's Pass, Canterbury.
Photo: J. W. Dawson.

The best known scree species is the penwiper plant (*Notothlaspi rosulatum*) (Fig. 112), which belongs to the cabbage family (Cruciferae). Its rosettes of grey, fleshy leaves closely overlap in a dome-like arrangement, which reminded early settlers of the similarly arranged and sewn together diamonds of felt on which they wiped their quill pens. The quite large flowers in massed heads are ivory in colour and have a very strong perfume reminiscent of stock in the same family.

Other notable plants restricted to mobile screes are *Ranunculus haastii* with large yellow flowers, *Wahlenbergia cartilaginea,* and *Lobelia roughii,* which has distinctive elk's-horn-like leaves with red teeth. *Stellaria roughii* in the chickweed genus is also common, as are two species of *Leptinella* of which *L. atrata*, with its almost black flowers, is the most remarkable.

The scree endemics die down in winter and most are perennials. The sole exception is the penwiper, which flowers in its second year then dies.

A number of species common on the more stable parts of screes particularly near the margins are also to be found in ordinary fellfield. Among these are *Epilobium rubromarginatum, Poa buchananii* (a grass) and patches of *Acaena glabra*. *Acaena glabra* belongs to the bidi-bid genus, but its seeds do not have the tiresome habit of attaching themselves to clothing or the wool of sheep, as do those of its more common lowland relatives. The seeds of *Acaena glabra* have spines but these lack hook-like recurved hairs at their tips. Perhaps the most unusual species in this group is *Craspedia incana,* a ghost-like plant with all its parts densely clothed in long, white, woolly hairs.

## Snow Banks

At higher altitudes in sheltered hollows, such as cirques or smaller depressions, snow will persist well into the growing season. Plants growing in such sites must be able to reproduce in just a few months, but they have certain advantages over the plants of the surrounding fellfield. They have more shelter from gales and, with reduced erosion on their concave slopes, a deeper soil which has a steady if cold supply of water from the gradually melting snow. The flowers of some of the snow bank plants may actually open beneath the melting snow. This is the case for two species of *Caltha* with their Ranunculus-like flowers, those of *C. obtusa* being white and *C. novae-zelandiae* yellow. Snowbank plants form a continuous sward, with the snow patch grass, *Chionochloa oreophila,* usually predominating. Among other prominent plants are *Ranunculus sericophyllus* with deeply divided leaves and relatively large yellow flowers; *Celmisia allanii,* with patch forming rosettes of fluffy pale grey to snow white leaves; the related but pale green *C. haastii*, and two mat forming species, which although flowering plants look remarkably like mosses — *Raoulia subulata* and *Drapetes lyallii.*

*Figure 113* Summit of the Old Man Range, Central Otago. The surface in the foreground and back to the tors is almost completely covered with cushion moorland. Photo: J. W. Dawson.

## Cushion Moorland[145]

Most New Zealand mountains are steep with narrow ridges and distinct peaks, but in central and north Otago in the south of the South Island there are a number of quite different block mountains, bounded by faults, with flat, plateau-like summits dotted with prominent rocky tors (Fig. 113). During the last glaciation when the winds were strong and much of the landscape was open with little plant cover, dust storms would have been frequent and sometimes quite deep deposits of dust, known as loess, would have built up on the hills and mountains. On the loess mantling the flat tops of the Otago mountains, cushion moorland

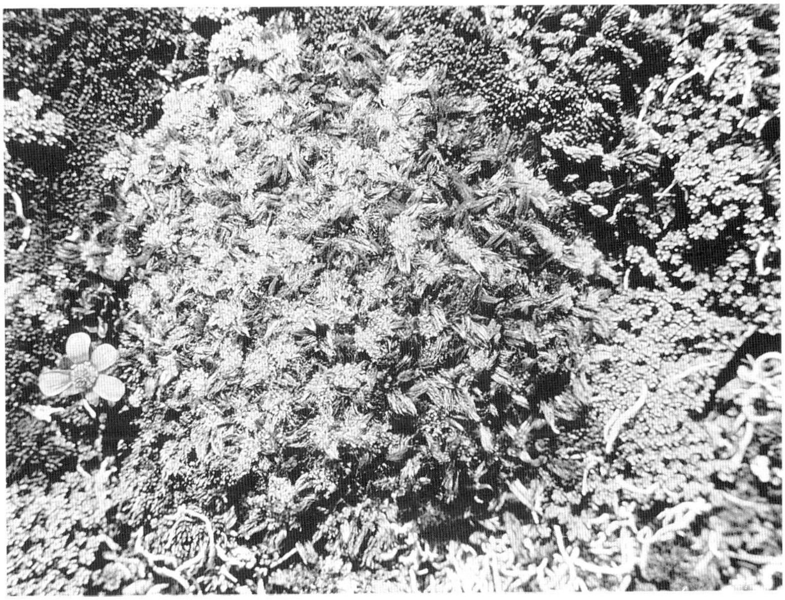

*Figure 114* Close up of cushion moorland. A silvery cushion of *Anisotome imbricata* surrounded by species of *Dracophyllum* and *Raoulia*.
Photo: J. W. Dawson.

forms a low but more or less continuous cover. Conditions here are at least as severe as those of the fellfield, with strong, cold gales from the south, heavy snow in winter and frequent severe frosts. Over a 5 year period on the Old Man Range at 1590 m there were on average 113 ice days, 179 freeze-thaw days and only 73 frost-free days per year.

In this type of vegetation small cushion plants are characteristic, with *Dracophyllum muscoides* the most common, but also the silvery *Raoulia hectorii, Anisotome imbricata* (Umbelliferae) with dense silver-grey hairs (Fig. 114) and *Phyllachne rubra*. Small rosette herbs grow in and between the cushions including *Anisotome lanuginosa* (also with dense matted silver-grey hairs obscuring the leaves) and the small spaniard *Aciphylla hectorii*. A few small grasses, sedges and luzulas of the rush family may also be present, and small grey lichens are often an important component.

As a result of freeze/thaw action the surface of the cushion moorland is often shaped on level areas into a very regular pattern of hummocks and hollows, which tend to elongate into alternating ridges and trenches on slopes.

CHAPTER 9

# Where Did Our Alpine Plants Come From?

Despite the fact that the alpine flora in New Zealand is considerably richer in species than the forest flora, its derivation poses the more puzzling problem. This is because geological evidence indicates that the present high mountains with their cold climates are relatively recent. In fact, for all except the last few of the past 100 million years it is suggested that New Zealand had warmer climates than now and much lower relief. In these circumstances it is difficult to imagine any habitats suitable for cold-tolerant plants that could have formed a nucleus for the present diverse alpine flora.

## Derivation from Plants Already Here

Cockayne suggested the possibility that a few alpine species from the much earlier mountain building period of about 100 million years ago (Rangitata Orogeny) may have survived through the warmer, mountainless times in suitable sites such as rocky bluffs.[63] He described these species as 'plastic', that is, that although requiring open habitats, they would have had a wide temperature tolerance. With the coming of the recent Ice Age and high mountains these surviving 'alpines' or their derivatives would have diversified into the new cold habitats.

Present rocky coast species of otherwise alpine genera could appear to support Cockayne's hypothesis. Some of these are: the spaniard, *Aciphylla squarrosa,* of the Cook Strait coasts; *Gingidia montana* on the Marlborough coast; *Celmisia mackaui* of Akaroa, Banks Peninsula and *C. lindsayi* of the south-east Otago coast; *Anisotome lyallii* of the south-east Otago coast, Fiordland and Stewart Island; and *Phormium cookianum,* perhaps equally common on coastal rocks as in the high mountains. The problem here is that these are all herbaceous flowering plants and 100 million years ago fossil evidence indicates that flowering plants, then in the early stages of their evolution, were predominantly woody. If the fossil record can be taken as a reliable guide (and some question this) it was not until long after the time of Rangitata Orogeny in higher northern (and perhaps also southern) latitudes that herbaceous flowering plants began to expand and diversify.

However, a modification of Cockayne's theory might be tenable. After herbaceous angiosperms became common some of them may have reached New Zealand by long distance dispersal from the northern hemisphere, and perhaps from a then warmer Antarctica, and, if tolerant of warmer temperatures, they may have established in various open lowland habitats. With the coming of alpine habitats these species would have been able to move into them and diversify.

This is not to say that the lowland species of herbaceous, predominately alpine genera in New Zealand are necessarily primitive and ancestral. During glacials, alpine vegetation probably extended at times right to the coasts, particularly in southern New Zealand. With each interglacial, forests would replace alpine plants in the lowlands, except on some exposed rocky coastal sites. Here some alpine species could persist and even give rise to new forms. In these circumstances we would have to regard the coastal species as derived from alpine species rather than as being their ancestral stock.

Wardle[108] suggested that in warmer pre-Ice Age times New Zealand may have extended much further south to the limits of what is now the submerged Campbell Plateau. Even in those warmer times he felt that such a southern extension would have had a cool temperate climate able to support a flora 'which would later give rise to the present mountain flora'. He based this idea on the present, fairly diverse, mostly herbaceous floras of the subantarctic islands, which share a number of genera with the New Zealand mountains, but which also have a number of endemic genera which, Wardle suggests, did not reach the main islands.

Unfortunately geologists do not support the idea of such a southern extension at that time, although even scattered islands to the south in warmer times may have provided adequate sites for the evolution of a stock of cool temperate plants.

In a later paper Wardle[146] proposed that a small element of the alpine flora may have been present in New Zealand throughout the warmer climates of the Tertiary, a view partly supported by fossil evidence. The generic and subgeneric groups concerned 'are small and taxonomically isolated, restricted to cool habitats, and either endemic to New Zealand or occurring elsewhere only in the Australian or American sectors of the South Temperate Zone'. Wardle thinks it unlikely that the species of this element have been derived from warm-climate Tertiary ancestors, or that they reached New Zealand by long distance dispersal. He lists 29 flowering species in 15 genera which are mostly confined to cool, wet, infertile soils. Of these, '*Nothofagus (N. menziesii* — silver beech) is a forest dominant, *Stilbocarpa* is a tall herb of far southern coasts, while the remainder are low growing plants ascending to subalpine or alpine levels, eight of them being more or less cushion or mat-forming'. Genera with species of the latter form are *Hectorella, Phyllachne, Donatia, Tetrachondra* and *Oreobolus*. He also suggests that certain shrubby, cold-tolerant, higher altitude conifers such as *Podocarpus nivalis, Lepidothamnus laxifolius* and *Phyllocladus aspleniifolius* var. *alpinus* may also have been present through the Tertiary rather than having been derived more recently from taller, lower altitude relatives. This view is based on a study which indicates that the shrubby forms are more primitive cytologically than the trees, and, if the cytological interpretation is valid, it follows that the former are unlikely to have been derived from the latter.

It had been earlier suggested[147] that this alpine element may have reached the New Zealand mountains from Antarctica, as presumably would also have been the case for the comparable elements in Tasmania and South America. For New Zealand this would imply a migration via the subantarctic islands, but on the whole Wardle does not favour this view as only 6 of the 18 genera concerned occur on these islands, and in the case of *Nothofagus* poor dispersal ability would suggest that crossing such wide ocean gaps would not have been possible. So if species of this element were already present in New Zealand before colder climates and mountains developed, where did they grow? Wardle points out that even today a number of the species of the group occur in the wet Westland climate within 300 m of sea level on infertile soils, and in Hawai'i, whose climate may be similar to that of New Zealand during the Tertiary, otherwise alpine species can be found in a bog at only 600 m altitude. 'If, during the period of Tertiary warmth (in New Zealand), there were infertile soils on peneplained uplands drenched by persistent mist and rain, some cool climate plants would undoubtedly exist . . . On the other hand, considerably colder conditions would seem necessary for the Tertiary existence of the high altitude genera *Hecto-*

*rella, Haastia, Phyllachne, Donatia* and *Rostkovia,* and it is perhaps in respect to these that the suggestion of immigration from Antarctica at the end of the Tertiary is most attractive.' Finally, Wardle suggests, the steep slopes with young soils of most of the present mountains would not have been suitable for this element, so it has remained as a minor component of the alpine flora as a whole.

The possibility of the derivation of alpine species from warm-climate forest ancestors in New Zealand has already been mentioned. Most authors would accept this as one source, but most also consider its contribution as relatively minor — involving mostly subalpine and alpine shrubs. For example, some botanists consider that divaricate shrubs are basically adaptations to Ice Age cold and drought and some of these, which belong to such genera as *Coprosma, Pittosporum, Melicytus, Myrsine, Aristotelia,* are to be found in subalpine and alpine habitats. In *Coprosma* and *Myrsine* there are even a few completely prostrate alpine species. Among non-divaricate derivatives, *Dracophyllum* has a number of needle-leaved and some broader-leaved shrubs in lower alpine and near cushion forms in higher alpine situations and *Hebe* is well represented by alpine shrubs, some of which are of whipcord form. *Astelia* is an herbaceous genus with a number of forest floor species and one epiphytic species, which is also quite well represented in the alpine zone.

## Immigrants from The Northern Hemisphere

However, if it is true that only a minor part of the New Zealand alpine flora is derived from plants that were already present here, where did the majority of the alpine species come from? Raven[148] suggests that many of them are derived from ancestors which reached New Zealand by long distance dispersal after the mountains had formed here. He believes that the ultimate source of the genera to which these ancestral colonisers belonged was temperate Asia and that they were only able to reach Australasia in recent geological times after mountains, contemporary with those of New Zealand, had formed in south east Asia, Indonesia, New Guinea and eastern Australia. Species of these genera 'mountain hopped' across the tropics to eastern Australia and then across to New Zealand by the agency of the strong, frequent, westerly winds of the 'Roaring Forties'. Raven explains the fact that some of the genera concerned have many more species in New Zealand than in eastern Australia by the higher, ecologically more diverse mountains in New Zealand providing a setting for extensive radiation of the immigrants after their arrival from Australia. In view of the direction and strength of the prevailing winds he regards migration from New Zealand to Australia as improbable.

Some of the genera that Raven believes reached New Zealand in this way are represented here by only one or a few species, for example, *Anemone, Geranium, Linum, Stellaria* and *Viola,* but others have dozens of species and include *Epilobium, Ranunculus, Myosotis,* the grass genus *Poa,* the sedge *Carex* and the rush genus *Juncus.* Other large genera are not known in north temperate regions, but at least some of their species were originally placed in northern genera from which they may have evolved within Australasia, for example, *Celmisia (Aster), Anisotome* and *Aciphylla (Ligusticum),* and *Hebe (Veronica).*

In a response to Raven's theory Wardle[149] claims that Raven underestimates the possibility of plants dispersing from New Zealand to Australia. He points out that vigorous easterly winds can flow from New Zealand to Australia when a tropical depression meets an anticyclone over New Zealand and also notes that there are several bird species migrating annually between New Zealand and Australia, which might serve as seed transporters. There is also the possibility of direct transport between New Zealand and northern Asia and north-west America by a number of migratory birds. A few mountain species of grasses, which are the same as or closely related to north temperate species, may have arrived in this way.

Wardle further states that it is necessary to know something about patterns of evolution within a genus before one can hypothesise about directions of migration. If for instance there were in a genus a mixture of primitive and advanced species in New Zealand, but only a few advanced species in Australia, then it is likely that the latter would have been derived from New Zealand.

The patterns in *Epilobium* indicate that Raven's interpretation could be correct with the exception of two species in Australia which may have been derived from New Zealand. However, since Wardle's article, abundant fossil pollen of *Epilobium* has been found in New Zealand of Oligocene age, a period long before the formation of the present mountains. So if *Epilobium* persisted in New Zealand until colder climates developed, then some at least of the present species may derive from that early stock, though others may have come from more recent colonisations.

In *Ranunculus* the affinities of the lowland and montane species are with Australia, but the alpines are most like species in South America suggesting a different origin.

Wardle suggests that an ultimate South American derivation is also possible for the snow grass genus *Chionochloa* , which exhibits cytological and other similarities with *Cortaderia* (which includes Pampas grass) of South America and New Zealand. Similarly most of the New Zealand species of *Plantago* belong to an otherwise South American section of the genus and so may have been derived from there rather than from temperate Asia via Australia.

## Southern Origins of Some Alpine Genera

I now want to consider some of the southern genera thought by some botanists to be related to and probably derived from north temperate genera. I propose an alternative hypothesis — that they may have evolved independently in the southern hemisphere.

With regard to the derivation of herbaceous genera in general it is worthwhile to consider what the vegetation pattern of the world may have been before this growth form suited to cooler, seasonal climates became widespread. The fossil evidence suggests that at these earlier times rain forest extended from the equator half way to the poles or even further and we can speculate that within such forests there would have been the woody ancestors of the herbaceous genera. Most of the latter would have originated in the northern hemisphere because of the greater area of land there and the more seasonal climates, but there seems no reason why other herbaceous genera could not have been derived in the southern hemisphere also from woody stock at the fringes of the great rain forest belt.

### New Zealand Umbelliferae

In an earlier review[150] I proposed two evolutionary lines within the more notable New Zealand genera of the main subfamily of the Umbelliferae (carrot family). One included *Aciphylla* (spaniards) and *Anisotome* and is distinguished by the inflorescences arising at the centre of leaf rosettes and terminating their growth and by their dioecism (separate male and female plants). The other group, comprising *Scandia,*

*Gingidia* and *Lignocarpa,* has mostly lateral inflorescences and is mostly gynodioecious (separate female and hermaphrodite plants). In the latter group I considered *Scandia* to be the most primitive genus as it is semi-woody with extended stems. *Scandia rosifolia* is the more shrubby of the two species of the genus and it is restricted to the northern half of the North Island and there often to mild coastal habitats. It is not difficult to imagine this species, and perhaps even woodier, now extinct, relatives, existing in New Zealand in warm pre-Ice Age times. From *Scandia* the completely herbaceous montane to alpine *Gingidia* could have evolved and also the specialised *Lignocarpa* of shingle slips. Since this idea was proposed a probable hybrid between *Aciphylla squarrosa* and *Gingidia montana* has been discovered in Marlborough[151] which suggests that the two lines are more closely related than I thought and that all the genera concerned may have been derived from woody ancestors with a long history in New Zealand. In eastern Australia there are two species assignable to *Aciphylla,* one to *Anisotome* and three to *Gingidia*, including a localised occurrence of *G. montana*. In view of the foregoing discussion it seems more likely to me that the Australian species have a New Zealand ancestry than vice versa.

In a recent review Webb[152] gives a largely opposite interpretation. He considers the few Australian species of these genera to be primitive, while New Zealand has a range from primitive to specialised forms. In contrast to my view, he regards the woodiness of *Scandia* as a recent specialisation. With one exception, he suggests that primitive species of these genera migrated from Australia to New Zealand with consequent diversification and specialisation here. The exception is *Gingidia montana,* where the localised occurrence in northern New South Wales he considers to be a case of recent migration from New Zealand to Australia.

Of course, at an earlier time before high mountains existed in New Zealand, there could have been only primitive species of these genera in both Australia and New Zealand and migration could have been in either direction. With the formation of the high mountains in New Zealand more specialised forms would then have evolved here.

According to yet another view some would argue that long distance migrations have not been involved at all. The occurrence of the same or related alpine plants in both Australia and New Zealand would derive from the time when they shared the same region before it was sundered by the drifting away of the New Zealand crustal complex from Australia.

With little direct evidence from the past we can only speculate.

### Celmisia

As well as the 60 or so species of *Celmisia* in New Zealand there are two related genera in the subantarctic islands: the semi-woody *Damnamenia* and the large, tufted, herbaceous *Pleurophyllum*.[153] *Pleurophyllum* is considered to be intermediate in some respects between *Celmisia* and one group of the woody genus *Olearia*, also strongly represented in New Zealand. This fact, plus possible wild hybrids between *Olearia* and *Celmisia*, suggests a derivation, perhaps within New Zealand, of *Celmisia* from *Olearia*. Some of the New Zealand celmisias are described as sub-shrubs, but the five species in east Australia are herbaceous and in Wardle's words 'seem to represent end points of evolutionary pathways stemming from New Zealand'.

### Hebe and Related Genera

New Zealand has perhaps 100 species of *Hebe* and there are only three species elsewhere. Two of these in southern South America are shared with New Zealand *(H. elliptica, H. salicifolia)* and the third is *Hebe rapensis* on Rapa Island in French Polynesia. The related but generally less woody *Parahebe* is also represented in eastern Australia and New Guinea and *Chionohebe*, probably derived from *Parahebe*, is a genus of small, high alpine cushion plants of which two of the New Zealand species extend to south-eastern Australia. *Veronica*, from which Raven[148] would consider the *Hebe* alliance to have been derived, is a mostly north temperate, herbaceous to slightly woody genus. It seems reduced and specialised and so is unlikely to have been ancestral to the woody hebes, some of which may become small trees with trunks sometimes 30 cm in diameter. There are differences too between the basic chromosome numbers of *Veronica* and the *Hebe* alliance which do not suggest a close relationship. There are indeed a few true veronicas in eastern Australia but, as Wardle[149] comments, 'it seems . . . possible that the handful of Australian true veronicas derive from recent northern immigrants that are not directly related to the *Hebe* alliance'.

Presumably *Veronica* was originally derived from woody, northern hemisphere ancestors, which may have been related to and perhaps also ancestral to the *Hebe* alliance.

Once again it is clear that much more research is required into a number of genera before we can attain any degree of certainty about the relative proportions of New Zealand alpines that have been derived from:

(a) geologically recent immigrants from the northern hemisphere;

(b) more ancient immigrants from the northern hemisphere; or

(c) genera of southern hemisphere origins.

CHAPTER 10

# Plants of the Outlying Islands

The small outlying islands are found at distances from mainland New Zealand of about 450 km (Aucklands) to about 1300 km (Lord Howe) (Fig. 2). Their botanical links with New Zealand range from significant to strong.

A group comprising Macquarie, the Aucklands, Campbell and Antipodes lies south of New Zealand and is conveniently termed the subantarctic islands; the Chatham Islands are temperate and lie to the east of the central South Island; while the Kermadecs, Norfolk and Lord Howe lie to the north of the North Island and can be termed subtropical. The Kermadecs and Macquarie lie on submarine oceanic ridges resulting from the interaction of crustal plates, but the other islands stand on submarine plateaus and ridges, which extend from New Zealand and are probably all submerged continental crust. Macquarie Island is comprised of serpentine and basalt pushed up from the sea floor, but the others are entirely or largely volcanic. One outcrop of granite is known from the Auckland Islands, some schist and sedimentary rocks occur on Campbell Island and the main Chatham Island has limited areas of schist and limestone in the north. The small granitic Bounty Islands have no land flora.

## The Subantarctic Islands

Climatically these islands are subjected to strong, often gale force winds, cool temperatures with no great difference between summer and winter, and rainfall of 1000–1500 mm per year. The latter is moderate, but as it falls on more than 300 days per year it is, in that sense, excessive. The frequent rain, infrequent sunshine, high humidity and coolness result in wet soil conditions and slow decay of dead plant parts. The latter persist for a long time at the bases of living plants and gradually form into peat which clothes the landscape to depths varying from a few centimetres to many metres. A special environmental factor, particularly in sites near the sea, is the presence of large mammals and birds, including sea elephants, sea lions, seals, penguins and albatrosses. Considerable damage, both physical and chemical, may be caused to plants by the activities of these animals, although a few species thrive on the soils enriched by their droppings. The majority of the plants of the subantarctic islands are herbaceous with large tussock grasses and robust herbs being particularly conspicuous. Some of these herbs are luxuriant in their growth with larger leaves, flower heads and often more colourful flowers than their mainland relatives.

*Figure 115* Leonard Cockayne and a large tussock of *Poa litorosa* on the Auckland Islands. Photo: S. Page.

## Macquarie Island[154,155]

Macquarie Island at 54°S is the furthest south and the chilliest of the outlying islands. It is long and narrow, about 35 km by 3–6 km, and although the highest point is only 433 m above sea level, it rises like a wall from the sea to a narrow plateau dotted with ice-formed lakes. In general form it is not unlike some of the narrow, flat-topped mountains of the Otago region. Geological evidence suggests that it was completely under ice during the last glaciation and it is generally believed that the present flora derives from long distance dispersal since that time. During the last glaciation however, sea level was 100 m or more lower than now so it is possible that some plants survived on land which is now below the sea.

On the raised beach terraces and lower cliff slopes the vegetation is mostly dominated by large tussocks of the grass *Poa foliosa* which develop trunk-like bases of sometimes a metre in height. *Stilbocarpa polaris,* which belongs to a distinctive genus restricted to the southern extremity of New Zealand and its subantarctic islands, is associated with the tussocks particularly in places enriched by animals. Although *Stilbocarpa* belongs to the mostly woody Araliaceae (including *Pseudopanax*) it is completely herbaceous with large soft leaves, after the style of rhubarb, with blades sometimes as much as half a metre across. The flower heads are also large with masses of small, yellowish green flowers. The upper cliff slopes are mostly occupied by a short tussock grassland in which *Pleurophyllum hookeri* may be a conspicuous component. *Pleurophyllum* is a genus restricted to the subantarctic islands, but related to both *Celmisia* and *Olearia.* The three species of *Pleurophyllum* are notable for rosettes up to a metre or more across of broad, longitudinally ribbed leaves, and for their colourful flower heads. In *P. hookeri* the flower heads, coloured reddish purple, are relatively small as their 'petals' are greatly reduced.

There is a sudden change of vegetation type on the windswept summit plateau; there is much exposed rock debris and the conspicuous plants are of cushion form. Some of these are mosses of the genus *Racomitrium*, but the most common cushion plant is *Azorella selago* of the Umbelliferae (carrot family). This is the only occurrence of the genus in the New Zealand region, but it and related genera are prominent in the Andes of South America where some species form huge cushions larger than our 'vegetable sheep'. *Azorella selago* also occurs at the southern tip of South America and on other islands scattered around the Antarctic Ocean. The *Azorella* cushions form a distinctive pattern well described by Gillham:[156] 'On leeward slopes with downhill winds *Azorella* plants coalesce laterally to give terraces as much as a mile long, the ribbon development arising from the prevention of both upward and downward growth by the wind. As many as 100 terraces, each about 1½ yards high, may run parallel to the contours to form a gigantic staircase.' The cushions grow directly on rock or glacial till and form little or no peat, suggesting an affinity with the high alpine cushion moorland of the flat-topped Otago mountains. Dwarfed versions of the *Pleurophyllum* and *Stilbocarpa* grow in the cushions as well as, surprisingly, the fern *Pyrrosia eleagnifolia*. The latter is often common as a sun epiphyte on trees in New Zealand and other species of the genus occupy similar habitats in the Asian tropics.

In all there is a modest total of 40 native species of vascular plants on Macquarie. One grass among these, *Deschampsia penicillata*,[157] is considered to be endemic although related to other subantarctic species, and four others are not found elsewhere in the New Zealand region, but are shared with several other higher latitude islands — *Azorella selago,* already discussed, *Acaena magellanica, Ranunculus biternatus* and *Poa cookii.* The first three also occur on the southern tip of South America.[158]

About six of the remaining species are shared only with the New Zealand mainland and the rest are scattered around the subantarctic zone including the New Zealand subantarctic islands, and in some cases the New Zealand mainland or both New Zealand and Australia.

The low endemism of the Macquarie flora and the wide distributions of most of the species indicate recent derivation. *Nothofagus* logs, determined anatomically as belonging to South American species, have been washed up on Macquarie, but after such a journey it is unlikely that any seeds would have survived in bark crevices.[159] Fern spores could have arrived by wind and seeds could have travelled internally or externally with either migratory sea birds or land bird stragglers.

## Auckland Islands

The Auckland Islands are currently credited with 187 native vascular plant species,[160,161] of which perhaps 6 are endemic and a further 22 are found elsewhere only on other subantarctic islands in the New Zealand region.

The relative richness of the Auckland Islands' flora compared with Macquarie reflects partly the lower latitude (51°S); partly the larger size of the islands (the main Auckland Island and the barely separated Adams Island are about 50 km long by 10-30 km wide); and partly a less severe climate during the last glaciation which probably allowed some of the previous interglacial flora to survive. The last is also likely to be true of the Campbell and Antipodes Islands.

The Auckland Islands also contrast topographically with Macquarie in having a very indented eastern coastline and, between the glacially steepened valleys, mostly gradual slopes leading up to ridges and distinct peaks, sometimes flat-topped, up to 668 m in altitude (Fig. 116).

*Figure 116* General view of the Auckland Islands from the south. There is a coastal fringe of southern rata forest (*Metrosideros umbellata*) and above that, shrubland with 'lanes', then alpine vegetation. Photo: P. N. Johnson.

## Rata Forest

The Aucklands are the only islands in the New Zealand subantarctic with forest. Forest forms a narrow zone up to about 50 m above sea level on sheltered shorelines and is dominated by southern rata (*Metrosideros umbellata*) in association with *Dracophyllum longifolium* var. *cockayneanum* and *Pseudopanax simplex*. Undershrubs are *Coprosma foetidissima* (stinkwood), the *Dracophyllum* and *Myrsine divaricata,* while on the forest floor the ferns *Polystichum vestitum, Asplenium scleroprium,* and *Blechnum durum* are abundant in places. *Fuchsia excorticata* has been recorded from one locality and the tree fern *Cyathea smithii* from several sheltered north-facing valley sides in the eastern inlets. The trees of this forest are unusual in that the lower parts of their trunks and branches are inclined or even prostrate as a result of wind action. Cockayne[162] describes this very graphically: 'Everywhere are the massive prostrate and semi-prostrate trunks of the southern rata, sometimes lying close to the ground, at other times forming great arches, or at others again natural bridges over the deep depressions of the forest floor.'

At a few localities at the north-east tip of the main island and on adjacent islets there is also a coastal forest of the large-leaved *Olearia lyallii*. This species has increased its range in Port Ross over the past century and may have reached the Auckland Islands from southern New Zealand as recently as early last century.

### Shrubland

Above the narrow forest zone there is a wider belt of 'scrub' or shrubland. Some of the scrub species are also found in the forests, but in the shrubland they are reduced to 1–2 m in height and when dominated by *Myrsine divaricata* they are so dense as to be almost impenetrable. *Myrsine divaricata* is the dominant species on the upper parts of the valley sides and *Dracophyllum longifolium* immediately above the rata forest. On some gradual slopes the scrub is arranged in an unusual pattern (Fig 116). It forms narrow strips, more or less parallel to the contours, which alternate with open lanes with a ground cover of low plants including hard cushions of *Oreobolus pectinatus,* stunted plants of the scrub species, gentians, orchids and scattered tussocks of *Chionochloa antarctica.*[163] We still await a detailed study which might lead to an explanation of this pattern.

### Tussock Grassland

Above 300–400 m in altitude the shrubland gives way to a grassland dominated by the large yellow-green tussocks of *Chionochloa antarctica*. It seems probable that a similar grassland originally occupied the 'lanes' within the shrub zone, but these have been greatly modified by the activities of wild pigs.

Several large herbs were a conspicuous feature of the tussock grassland particularly at lower elevations and this is still the case in 'Fairchild's Garden' near sea level on Adams Island, where pigs are not present. Cockayne[162] describes the area as 'a wonderful collection of stately herbs with immense leaves, and in some cases masses of showy flowers'.

*Anisotome latifolia* has large, dark green dissected leaves and broad, pink to purple flower heads. *Stilbocarpa polaris,* already described for Macquarie, may also be common. Two species of *Pleurophyllum* are conspicuous — *P. speciosum* (Fig 117) with rosettes of almost circular, pale green longitudinally pleated leaves up to 30 cm in diameter and *P. criniferum* with even longer, but ascending, leaves up to 1 m long. The former has quite large flowers with purple centres and pinkish 'petals', while the latter, although its flower heads may be 1.5 m high, has red, dome-shaped button flowers similar to those of *Pleurophyllum hookeri.* Towards the upper limits of the garden *Pleurophyllum speciosum* and *Bulbinella rossii* become particularly abundant. *Bulbinella rossii* is much larger than its mainland relatives and has broad arching leaves and large orange-yellow flower heads.

*Figure 117* Close up of large herbs on the Auckland Islands. The plants in flower with broad ribbed leaves are *Pleurophyllum speciosum.* The dissected leaves on the left are *Anisotome latifolia* and the rounded leaves with toothed margins centre and right are *Stilbocarpa polaris.* A flower head of the latter appears in the left background. Photo: D. J. Campbell.

### Bogs

At higher altitudes, and also in the 'lanes' at lower altitudes, on very wet, more or less level sites there is a low plant cover of mat or cushion plants including *Carpha alpina, Astelia linearis, Oreobolus pectinatus, Centrolepis ciliata* and 'glistening green mats' of *Damnamenia vernicosa,* an endemic genus formerly included in *Celmisia.*

### Fellfield

Some of the exposed summit peaks have a stony mineral soil with little peat development. They support a variety of plants, including *Pleurophyllum hookeri* which is often common, *Plantago aucklandica* with its large leaf rosettes, *Ranunculus pinguis* with unusually large yellow flowers for the size of the plants and *Myosotis capitata* with brilliant dark blue flowers.

## Campbell Island

Campbell Island is found at 51.5°S. Topographically it is similar to the Auckland Islands but is smaller, measuring about 16 km by 16 km. It also has fewer vascular species, 128, of which the majority are shared with the Aucklands.[164] Four species are considered to be endemic.

It is a little difficult to reconstruct the original vegetation patterns as the island was operated as a sheep station from late last century until 1931 and as a result some palatable species became rare while unpalatable species became more common. A fence was constructed across the narrow centre of the island in 1970 and all sheep removed from the northern half of the island. It is encouraging to learn that some species are now recovering strongly north of the fence.[165]

Although not much further south than the Auckland Islands there is no forest fringe on Campbell Island. Instead up to about 200 m altitude in sheltered places there is a dense shrubland comprising *Dracophyllum scoparium* and *D. longifolium.* Sometimes two species of small-leaved coprosmas and *Myrsine divaricata* can be found mixed with the *Dracophyllum* and also scattered through the tussock grassland.

Cockayne[162] recognised two zones in the tussock grassland, the lower dominated by *Poa litorosa* and the upper by *Chionochloa antarctica.* Scattered throughout are several species of herbs and small shrubs, including the two larger species of *Pleurophyllum, Bulbinella rossii, Gentiana antarctica* and the blue-flowered *Hebe benthamii.*

The plants of cushion bogs and rocky mountain tops are mostly the same as those of comparable habitats in the Auckland Islands.

## Antipodes Island

At 49°S Antipodes Island is further north than the other vegetated subantarctic islands, but being only 7 km by 5 km is much the smallest and has a small flora — about 66 species of which 4 are endemic.[166] Above the fringing cliffs the island slopes gradually to a high point of about 400 m. The vegetation is almost entirely tussock grassland dominated by *Poa litorosa*. Near the shore the tussocks are particularly large and in Cockayne's words they 'grow so closely together out of the wet peaty soil that it is hardly feasible to force a passage between them, and it is much more easy to walk upon their tops, stepping from tussock to tussock'.[162] Further upslope the fern *Polystichum vestitum* in its semi-tree fern form becomes an important component of the grassland, but there is also 'much bright-green *Anisotome antipoda,* pale bluish-green *Acaena antarctica* (climbing over the tussocks), and the tender green fern *Histiopteris incisa*'. In sheltered places *Stilbocarpa polaris* may form dense stands. Where ground has been manured by the giant petrel the endemic herb *Senecio antipodus* is abundant.

Shrub patches of *Coprosma antipoda* occur in sheltered gullies.

Localised bogs are largely devoid of tussock grass, but have an abundance of *Anisotome antipoda, Stilbocarpa polaris* and patches of *Coprosma perpusilla* and the filmy fern *Hymenophyllum multifidum.*

## General Comment on the New Zealand Subantarctic Flora

From this review of the plants of New Zealand's subantarctic islands it is clear that their floras are quite closely related. In total about 200 species are currently recognised. In an earlier review Cheeseman[167] estimated that 27.5 per cent of the total New Zealand subantarctic flora was endemic, 69 per cent shared with New Zealand and in some cases other lands, and 3.5 per cent with other circumpolar islands and/or southern South America, but not New Zealand.

The New Zealand mainland and subantarctic island floras are closely related, but the significant level of endemism of the latter suggests that it has had a relatively long history.

## Chatham Islands

These islands lie about 820 km east of Banks Peninsula at 44°S. The main island is about 45 km north to south with a narrow waist 8 km across, widening to about 50 km in the north and to about 25 km in the south. The island is generally low lying with the highest point in the south only 284 m. About one quarter of the island's area is taken up by the Te Whanga lagoon and there are a number of other smaller lakes. Pitt Island, 16 km by 8 km, a little to the south, is the only other island of any size.

The Chatham Islands are a little greater in area than the Aucklands and lie in milder latitudes, so they have a larger vascular flora of about 320 species with a higher proportion of woody plants. At least 29 of the species and 8 susbspecific taxa are endemic and the great majority of the remainder are shared with New Zealand.

Rainfall on the main island is moderate, ranging from a minimum of 600 mm in the lowlands to probably over 1200 mm on the southern table land. Cool south-west winds predominate, atmospheric humidity is high, skies are often cloudy and temperatures are relatively cool for the latitude. Frosts, however, are rare and a few droughts exceeding 15 days can be expected each year. It is not a climate that one would expect to lead to the extensive development of peat, but in fact 60 per cent of the land surface is covered by peat and peat soils. Most of the peat is in raised bogs maintained by rainwater and hence is of low fertility. Wright,[168] who surveyed the soils of Chatham Island, suggests that a high level of salt deposition from strong winds may inhibit decay of plant litter and so encourage peat formation on the many flat or gently sloping sites.

Again there are some difficulties in reconstructing the original vegetation as a long period of Maori occupation was followed by European settlers with their attendant sheep and cattle. Early botanical accounts, especially that of Cockayne[169] early this century, are helpful in determining the nature of the original vegetation cover. At these earlier times, although there were still extensive forests, a number of the hillslopes and ridges at lower elevations and some ridges at higher elevations were occupied by a low plant cover dominated by bracken fern. As on the mainland this community was probably induced by Maori fire.

Recently Kelly[170] and Given and Williams[171] have surveyed the remaining areas of native vegetation of the Chathams and the following is largely based on their accounts.

**Broadleaf Forests**

The forests of the Chatham Islands are relatively low as none of the taller trees of the New Zealand mainland are present. There are no conifers at all, no beeches (*Nothofagus*), nor any of the other canopy dominants such as tawa (*Beilschmiedia tawa*), kamahi (*Weinmannia racemosa*), hinau (*Elaeocarpus dentatus*), or northern or southern rata (*Metrosideros robusta, M. umbellata*). In more open sites a notable absentee is the cabbage tree (*Cordyline australis*).

On steeper slopes, too well drained to permit the development of peat, the forest is relatively rich in species. A close canopy at 6 to 13 m is dominated by endemic species: *Coprosma chathamica* (the largest species of the genus), *Pseudopanax chathamicus, Melicytus chathamicus, Myrsine chathamica,* the striking yellow-flowered *Brachyglottis (Senecio) huntii*, and on more fertile soil the nikau palm, *Rhopalostylis sapida*.

Among the few understory species are kawakawa (*Macropiper excelsum*) and tree ferns, particularly *Dicksonia squarrosa* and *D. fibrosa*. The common vines are supplejack (*Ripogonum scandens*) and *Muehlenbeckia australis*. A number of epiphytes are present including the fern *Pyrrosia eleagnifolia* and the orchids *Earina mucronata* and *E. autumnalis,* but there are no asteliad nests.[172] This type of forest is now restricted to gullies in the south of the main island (Fig. 118) and the southern part of Pitt Island.

On more or less level sites in the lowlands, karaka (*Corynocarpus laevigatus*) dominated where drainage was good, giving way to *Plagianthus regius* var. *chathamicus* on the most fertile soils and to *Sophora microphylla* on calcareous sediments. On coastal cliffs and some dune country *Olearia traversii* played a similar role to that of pohutukawa (*Metrosideros excelsa*) in the northern North Island. In swampy situations there was a distinctive forest of this species in combination with *Coprosma chathamica*. These forest types have been almost entirely cleared for farming.

Plant communities associated with bogs (Fig. 119) are restricted to the main island and these are mostly found on gentle slopes at higher elevations where rainfall is high and mists frequent.

### Bog Forest

Bog Forest is dominated by *Dracophyllum arboreum* and was once very extensive. It is now largely restricted to the southern tableland. The *Dracophyllum* has quite broad leaves when young, but the adult leaves are needle-like. This species often begins life as an epiphyte on a tree fern and eventually becomes free standing on its own coalesced roots. Of *Dracophyllum arboreum* Wright[168] says: 'In many respects a parallel can be drawn between the kauri tree (*Agathis australis*) of Northland . . . and *Dracophyllum arboreum* in the Chathams. Both trees accumulate a large mound of organic residues, both condition strongly the soil leaching processes and bring about podsolisation, and both become the dominant species when the soils are sufficiently conditioned. The litter in the case of each species consists of leaf, twig and bark residues of high resinous composition.' The tree ferns *Dicksonia squarrosa* and *D. fibrosa* are prominent in the understory and filmy ferns on the forest floor and, as epiphytes, are a prominent feature.

### Shrub-Rushland

In wetter places areas of this type of vegetation formed a mosaic with bog forest and were known as 'clears' by the settlers. The dominant species is the bamboo-rush, *Sporadanthus traversii,* which forms a deep and fibrous peat. This genus with its single species is restricted to New Zealand and despite the inroads of farming is still to be found locally in the Waikato Basin of the North Island as well as the Chathams.

In the wettest hollows of the bogs sphagnum moss predominates and shrubs of *Dracophyllum scoparium* and the purple-flowered *Olearia semidentata* are scattered throughout. Most of the 'clears' have been completely altered by repeated burning.

A low shrubbery dominated by the endemic *Styphelia robusta* occupies dry ridges in the peatlands of the southern tableland.

### Coastal Communities

A full range of coastal habitats, from sand dunes to rocky cliffs, can be found in the Chatham Islands. A number of the species are the familiar plants of such sites on the New Zealand mainland, but there are also some notable endemics. Among these, certain large herbs once formed a band of lush vegetation at the top of many sandy and stony storm

*Figure 118 (opposite above)* Chatham Island. Southern cliffs with low forest. Photo: D. R. Given.

*Figure 119 (opposite below)* Chatham Island. Bog scene on the southern tableland. Photo: D. R. Given.

beaches: the nettle *Urtica australis, Embergeria* (*Sonchus*) *grandifolius* and *Myosotidium hortensia. M. hortensia,* now widely cultivated, could be called a giant forget-me-not. The leaves, like those of *Stilbocarpa* in the subantarctic, are rhubarb-like and almost fleshy and the flowers in large heads are blue in the centre grading outwards to white. It is a plant stock find very palatable and is now quite rare. Today's sand dunes are dominated by the introduced marram grass.

Endemics of sea cliffs include *Hebe chathamica, H. dieffenbachii* and the grasses *Poa chathamica* and *Festuca coxii. Myosotidium hortensia* may still be found in inaccessible places. Dense shrubberies of *Olearia chathamica* may be prominent along cliff tops. A large, grey-green, remarkably soft-leaved *Aciphylla, A. dieffenbachii,* was formerly common along the southern cliffs of the main island and is still frequent on Pitt Island. A smaller species, *A. traversii,* is a plant of tableland bogs, but it, too, has been reduced by stock.

In highly fertile peaty places riddled with muttonbird burrows, grow luxuriant masses of *Carex trifida* and the endemics *Leptinella featherstonii* and *Senecio radiolatus*. Such sites are now confined to the smaller islands.

With their absence of groups of poor dispersal ability, particularly conifers, the Chathams have an isolated oceanic rather than a continental flora. New Zealand seems to be the principal source area but, as with the subantarctic islands, the presence of a significant element of distinctive endemics indicates quite a long period in isolation.

## Subtropical Islands

### Kermadec Islands

The Kermadecs are a row of small islands about 1000 km north-east of New Zealand. The largest and northernmost island is Raoul at 29°S, but it measures only about 10 km by 8 km with a highest point of 520 m. It is a volcano which is active from time to time. Macauley Island, the next island to the south, is only 2 km long, and the southernmost island with a significant flora, Curtis, is even smaller.

With the warm, humid climate, forest is the predominant vegetation cover.[173] On Raoul Island Oliver[174] distinguished a lower altitude 'dry' forest and above 2–300 m a higher altitude 'wet' forest. *Metrosideros kermadecensis,* up to 20 m tall, dominates the canopy of the lower forest. Karakas (*Corynocarpus laevigatus*) and ngaios (*Myoporum kermadecense*, closely related to *M. laetum*) also reach 20 m in height in sheltered places; considerably greater dimensions than they usually attain in New Zealand. There is a rather sparse subcanopy dominated by *Myrsine kermadecensis* but also with mahoe (*Melicytus ramiflorus*), ngaio, karaka, *Melicope ternata, Coprosma acutifolia,* the nikau palm relative *Rhopalostylis baueri* var. *cheesemanii* (sometimes forming picturesque groves), *Cyathea milnei*, and the endemic *Homalanthus polyandrus*, now increasing markedly owing to the great reduction in numbers of goats. The fern *Pteris comans,* up to 2 m high, is the commonest plant of the forest floor (although in many places the introduced aroid *Alocasia macrorhiza* is dominant), but there are also a number of other ferns and the large leaved form of kawakawa (*Macropiper excelsum* f. *psittacorum*) along the north side.

The upper or 'wet' forest is mostly about 10 m high and comprises the same species as the lower forest with the addition of *Ascarina lucida* var. *lanceolata,* which dominates the understorey, and, in a few places, the tree fern *Cyathea kermadecensis. C. kermadecensis* occasionally forms impressive groves where some specimens, according to Oliver, have trunk bases 1–2 m in diameter. The forest floor and trunks and branches of trees are covered with mosses and ferns. Common epiphytes familiar to New Zealanders are the ferns *Phymatosorus diversifolius, Pyrrosia eleagnifolia, Asplenium flaccidum* and *Hymenophyllam demissum.* Much rarer are *Lycopodium varium, Tmesipteris lanceolata* and *Trichomanes venosum.* There are no lianes in the Kermadec forests and no large epiphytes apart from the occasional *Pseudopanax arboreus* var. *kermadecensis* and *Metrosideros kermadecensis.* The only other flowering plant epiphyte is *Peperomia urvilleana.*

Close to the sea in many places, as in New Zealand, is a dense shrubbery of the ngaio relative *Myoporum kermadecense.* Probably ngaio shrubbery was the dominant vegetation on Macauley Island, but the fires of early whalers and the activities of goats have almost eliminated the woody vegetation of that island.

In open, mostly rocky coastal habitats are a number of herbaceous plants, including a few ferns. Most of these grow in similar sites in New Zealand. *Coprosma petiolata* is a common endemic coastal shrub.

The total vascular flora of the Kermadecs is 113 species. Eighteen species and three varieties are considered to be endemic and those with New Zealand affinities are *Coprosma petiolata* and *C. acutifolia,* the tree ferns *Cyathea milnei* and *C. kermadecensis, Pseudopanax arboreus* var. *kermadecensis* (close to the New Zealand variety and like it sometimes a tree fern epiphyte), *Metrosideros kermadecensis*, (known as the Kermadec pohutukawa in cultivation in New Zealand; it has shorter, more rounded leaves than pohutukawa (*M. excelsa*)), *Myoporum kermadecense, Hebe breviracemosa* (now probably extinct) and the palm *Rhopalostylis baueri* var. *cheesemanii.*

About 80 of the 113 Kermadec species are shared with New Zealand, but as many of them are also to be found elsewhere in the Pacific and in eastern Australia, they have not necessarily all been derived from New Zealand.

The Kermadec Islands have arisen as volcanoes in isolation and it seems likely that their flora has resulted from long distance dispersal, with New Zealand probably the most important source.

## Norfolk Island[175,176,180,181]

At the same latitude as Raoul Island, but 1400 km to the west, Norfolk Island is equally small with the even smaller Phillip Island a little to the south. The two islands are remnants of a long extinct volcano. Norfolk is fringed by cliffs and has a plateau-like surface rising gradually to a high point only 350 m above sea level. Little of the original vegetation remains as it was largely removed early last century as a result of a penal settlement being established and the subsequent settling of the island by the descendants of the 'Bounty' mutineers from Pitcairn in 1856. It has been prevented from re-establishing since that time by fires and grazing of stock. The climate, although a little drier than the Kermadecs, is suitable for forest and records of early explorers show that the island was originally covered with dense forest. The most notable species of the flora is the so-called Norfolk Island pine (*Araucaria heterophylla*), the only conifer native to any of the outlying islands. This has survived rather better than some of the other species and is still a distinctive feature of the landscape. In the original forest the araucarias would have been emergent above the main canopy and may also have formed pure groves on less fertile sites as is the case with some of the araucarias in New Caledonia.

The lower strata of the forest included species related to or the same as New Zealand species: *Nestegis apetala,* the palm *Rhopalostylis baueri* var. *baueri,* the tree fern *Cyathea robusta, Melicytus ramiflorus, Macropiper excelsum* f. *psittacorum, Myoporum obscurum, Cordyline obtecta, Pennantia endlicheri* and species of *Meryta*. Unlike the Kermadecs, Norfolk Island has locally abundant lianes which include *Freycinetia baueri* var. *baueri* and *Muehlenbeckia australis,* shared with New Zealand, and species of *Clematis, Passiflora, Ripogonum* and several other genera. Vascular epiphytes also are more numerous — species of the orchid genera *Dendrobium* and *Bulbophyllum,* and species of *Peperomia* and among ferns *Asplenium australasicum, A. polyodon, Pyrrosia confluens* and the fern ally *Tmesipteris norfolkensis.*

The cliffs are occupied mostly by species widespread in the southwest Pacific, with the exception of *Phormium tenax,* shared with New Zealand, and *Coprosma baueri,* which occupies similar sites to its very close relatives *C. repens* in New Zealand and *C. petiolata* in the Kermadecs.

With about 174 species Norfolk has a rather larger flora than the Kermadecs. Of these perhaps 50 are endemic.

About 76 species are shared with New Zealand and, often, elsewhere.

## Lord Howe Island[177-181]

Lying about 850 km west of Norfolk at 31°S Lord Howe Island is much closer to Australia (600 km) than New Zealand (1300 km). The island is crescent shaped and about 10 km long by 3 km at its widest. Like Norfolk it is the remnant of a long extinct volcano, but it is much more mountainous and has much of its original vegetation largely intact. The northern half of the island includes groups of hills, but in the south are two steep mountains — Mt. Gower (875 m) and Mt. Lidgbird (763 m) (Fig. 120). Mild to warm temperatures and ample rainfall support dense forest over most of the island, with shrub associations near the sea on cliffs, and on the mountain tops.

Up to about 300 m there is a relatively tall forest with scattered emergent trees to 20 m in height. In some places and particularly at lower elevations the emergents are of the endemic *Ficus columnaris,* one of the species of the fig genus known as banyans. In these the spreading branches are supported by innumerable trunk-like 'column roots'. Elsewhere and tending to be at higher elevations the emergents are *Cleistocalyx fullageri,* which belongs to the fleshy-fruited subfamily of the family Myrtaceae. The main canopy of the lowland forest is about 15 m high and composed of a mixture of species of genera not found in New Zealand. Conspicuous are some of the handsome palms for which Lord Howe is noted — *Howea forsteriana* and *H. belmoreana* at lower elevations with *Hedescepe canterburyana* higher up. Both genera are endemic. Among the subcanopy trees and shrubs New Zealand relatives are more frequent

*Figure 120 (opposite)* Lord Howe Island. Mt. Lidgbird with *Howea* palms in the foreground. Photo: G. W. Gibbs.

— *Coprosma putida, Geniostoma petiolosum, Sophora howinsula, Myrsine platystigma, Myoporum insulare, Dysoxylum pachyphyllum, Macropiper excelsum* and *Bubbia howeana*. *Bubbia* belongs to a primitive flowering family, Winteraceae, which also includes the New Zealand *Pseudowintera*. Ferns, sedges and grasses are common on the forest floor. Lianes are abundant, but, except for a *Clematis,* belong to genera not represented in New Zealand. Epiphytes are infrequent.

Above 300 m altitude there is a change to a montane low forest without emergents. A number of the lower altitude species are present here with the addition of *Dracophyllum fitzgeraldii, Metrosideros nervulosa* and *Olearia ballii*. The Howea palms are replaced by *Hedyscepe canterburyana*. Lianes and epiphytes are uncommon in this forest.

On the mountain summits is a low forest or shrubbery 3–4 m high in which tree ferns, *Dracophyllum fitzgeraldii* and the palms *Hedyscepe canterburyana* and *Lepidorrhachis mooreana* are conspicuous. Other species from lower altitudes are also present. Higher altitude species are *Negria rhabdothamnoides* (an endemic genus related to *Rhabdothamnus* of New Zealand), *Corokia carpodetoides, Coprosma lanceolaris,* and in more open places *Leptospermum flavescens* and *Gahnia xanthocarpa*. Ferns are abundant on the ground, and epiphytic ferns, mosses and lichens on the branches of the trees and shrubs.

At most places along the coast there is a narrow zone of shrubs tolerant of salt spray. A number of species are present, but only *Coprosma prisca* provides a link with New Zealand.

With about 230 vascular species Lord Howe has the richest of the subtropical floras resulting no doubt from its more varied topography and nearness to Australia. About 75 species or a bit less than one third are considered to be endemic. Included among them is *Carmichaelia exsul*, the only member of the genus occurring outside New Zealand.

About 70 species are shared by New Zealand and Lord Howe, but most of these occur in a number of other places as well.

## General Comments on the Floras of the Subtropical Islands

The relatively small total number of species and the few not very distinct endemics indicate that the Kermadecs, like Macquarie, have gained their flora by long distance dispersal in fairly recent geological times. By contrast the larger floras of Norfolk and Lord Howe with their larger numbers of mostly distinctive endemics, including some endemic genera, suggest a much longer history with the possibility of land connections as well as long distance dispersal being involved. Both islands stand on submarine ridges of continental crust and there could have been a succession of volcanic islands in their vicinity, allowing floral continuity over a long period of time, perhaps from when the ridges were still emergent. Thus related species in New Zealand and, say, Lord Howe Island may have resulted from long distance dispersal in either direction at various times followed by divergent evolution, or from two populations of an ancestral form separated following submergence of the Lord Howe Rise. The suggestion of some sort of long term land continuity at the sites of Lord Howe and Norfolk is supported by the presence on each of these islands of at least one species belonging to groups considered to have poor dispersal ability: on Norfolk the conifer *Araucaria heterophylla,* related to but distinct from *A. columnaris* of New Caledonia; and on Lord Howe *Bubbia howeana* belonging to the primitive flowering family Winteraceae whose members are also considered to be unlikely to cross wide ocean gaps.

At least some species shared by New Zealand and the subtropical islands have presumably achieved their widely disjunct distributions more recently by long distance dispersal.

CHAPTER 11

# The Plants of other Southern Lands

In this chapter, as a necessary background to the consideration of the origins and history of New Zealand plants in the last chapter, we will review fairly briefly the plant coverings of certain southern islands and continents which have strong botanical links with New Zealand. The Hawaiian Islands, although north of the equator, will also be included as much of their flora appears to have been derived from the south-west Pacific.

## Australia

Most of Australia is much drier than New Zealand and as a consequence most Australian vegetation types and their floras are very different from ours.[182,183] Vegetation types with counterparts in New Zealand — moist closed forests and open alpine communities — are restricted to the wetter and more mountainous eastern fringe of the continent and Tasmania. Even here it is estimated that prior to the arrival of man only about 1 per cent of Australia was occupied by closed forests and much less than that by alpine communities.

### Closed Forests

These extend in scattered patches from north Queensland to Victoria and Tasmania. The closed forests without *Nothofagus* have been divided latitudinally into tropical, subtropical and warm temperate. The first is found in north Queensland and is characterised by leaves of mesophyll size; the second ranges from central Queensland to central New South Wales and has a predominance of notophylls (small mesophyll); and the last straddles southern New South Wales and eastern Victoria and is predominantly microphyllous.[184] A number of the genera prominent in these forests are shared with New Zealand, for example *Elaeocarpus, Beilschmiedia, Dysoxylum, Syzygium* and the vines *Freycinetia, Ripogonum* and *Parsonsia,* but in most cases there are many more Australian than New Zealand species. Woody vines and vascular epiphytes are conspicuous in the more northern of these Australian forests, but are less common in the south at the same latitudes as New Zealand's North Island. In Queensland 'strangling' figs may occur as emergents and, in places, species of *Agathis* and the related *Araucaria.* In Queensland also the abundant nest epiphytes are ferns — species of *Asplenium, Drynaria* and *Platycerium* (Stags Horn Fern) — equivalent ecologically to the asteliad nests of New Zealand.

Closed forests where species of *Nothofagus* dominate or co-dominate are scattered through south-east Australia near the coast and Tasmania. In eastern Victoria and Tasmania the species involved, *Nothofagus cunninghamii,* is closely related to *N. menziesii* of New Zealand. *Nothofagus moorei* of montane sites in north central New South Wales and near the New South Wales, Queensland border also belongs to the '*N. menziesii* group', but has larger leaves than either *N. cunninghamii* or *N. menziesii.*

As in New Zealand the *Nothofagus* forests in Australia have relatively few species and few or no lianes and vascular epiphytes. Several other tree species are often associated with *Nothofagus* including *Ceratopetalum apetalum* (perhaps the counterpart of *Weinmannia* of the same family in New Zealand) *Atherosperma moschata* and in Tasmania, the conifers *Dacrydium franklinii* and *Phyllocladus aspleniifolius.* Tree ferns are common, particularly in canopy gaps and gulleys.

### Open Forest

In this category trees grow sufficiently far apart so their crowns do not form a continuous canopy and the ground as a consequence is not heavily shaded.

In tall open forest, also known as wet sclerophyll, very tall species of *Eucalyptus* are dominant. One of these, *E. regnans,* is the tallest flowering plant in the world, attaining a height of 100 m or more. Tall open forest is restricted to maritime south-east Australia, Tasmania, and a small area at the tip of south-west Australia. Where soils are relatively fertile, ferns predominate in the ground cover and where it is infertile, heath-like shrubs, belonging to the Myrtaceae, Proteaceae, Rutaceae and other families predominate. This forest type is restricted to fairly moist climates and perhaps in the absence of fire it would be succeeded by closed forest.

Low open forest occupies drier sites, sometimes further inland than those of tall open forest, in eastern, northern and south-eastern Australia and Tasmania. Relatively short species of *Eucalyptus* predominate in the tree stratum together with species of other genera in the same family and other drought-resistant species such as the scale-leaved Casuarinas. On fertile sites, tall grasses form the ground cover and on infertile sites heath-like shrubs. Sometimes trees are absent and pure grassland or heathland results.

### Tall Shrubland

In the southern parts of South and Western Australia there is a special type of shrubland known as Mallee. Here the dominant eucalypts are multistemmed and regenerate after fire from bulky woody tubers known as lignotubers. Generally inland from the eucalyptus-dominated communities, where climates are drier and soils infertile, drought-resistant species of *Acacia,* particularly *A. aneura* (mulga), prevail. The sometimes sparse ground cover comprises grasses and shrubs.

### Hummock Grassland

Succeeding the Acacia shrubland in the arid heart of Australia is a grassland dominated by the prickly small tussocks of grasses, mainly belonging to the genus *Triodia.*

### Chenopod Low Shrubland

In south central Australia on calcareous, sometimes semi-saline, soils there is a low shrubland where the salt-tolerant family Chenopodiaceae is strongly represented.

### Subalpine and Alpine Communities[185]

Despite its size, the continent of Australia has generally low relief and it is only in the south-east and in Tasmania that mountains rise above treeline. The mountains are not of alpine form, but are basically raised plateaus with fringing scarps. Glaciation was extensive during the Ice Age in Tasmania and as a result there are more steep and rocky sites in the mountains there than on the mainland.

The zone 400–500 m below treeline is occupied by woodland dominated by a few small species of *Eucalyptus* (Snowgums). In Tasmania there are, in addition, thickets of a shrubby deciduous species of *Nothofagus, N. gunnii,* which belongs to the *N. fusca* group. Associated with the *Nothofagus* are species of *Athrotaxis,* the only southern hemisphere genus of the conifer family Taxodiaceae.

Frosty valley floors below and above treeline are occupied by a tussock grassland with species of *Poa, Danthonia* and other genera.

Above treeline, particularly on the mainland, an important plant community is a herbfield of short grasses and such genera as *Celmisia, Craspedia* and *Euphrasia* growing on fairly deep, well-drained soils. Below persistent snow patches is an even shorter herbfield with species of *Plantago, Neopaxia, Caltha* and *Ranunculus* as in New Zealand. Less common on the mainland, but well developed in Tasmania is a shrubbery on rocky sites comprising genera shared with the heaths of the lowlands.

Poorly drained sites support bogs with *Sphagnum* moss, a variety of small shrubs, sedges and small species of *Astelia.* In Tasmania alone cushion bogs with species the same as or related to those of New Zealand are frequent. Species present include *Donatia novae-zelandiae* and *Phyllachne colensoi* (shared with New Zealand) and species of *Oreobolus* and *Astelia.*

Above snow patches comes a type of fellfield with low species of *Coprosma* and *Colobanthus* and, on wind-exposed rounded ridge crests, a different type with, among other genera, *Chionohebe* and *Drapetes,* which are also shared with New Zealand. From photographs the latter community looks not unlike the summit plateau fellfield of Macquarie Island.

## New Guinea[186,187]

Narrowly separated to the north, the island of New Guinea contrasts with most of Australia in many respects — it has a moist tropical climate leading to a predominance of closed forests. Except for extensive plains in the south, it is much more mountainous and the mountains are steep and high enough, even at such low latitudes, to rise above treeline.

### Lowland Zone

Along estuaries mangrove forests are well developed and contain many species. Further inland on the plains, swamp forests are widespread, and in the south, where there is a distinct dry season, monsoon forests characterised by a proportion of deciduous trees occupy better drained sites. In the driest areas fire is an important factor and an open forest of Australian aspect and relationships covers the landscape. Species of *Eucalyptus* and other genera of the Myrtaceae dominate in these forests. However, the most widespread forest type up to 700 m is a species-rich, evergreen rain forest. Strangling figs may occur as emergents over a main canopy of buttressed trees up to 45 m in height. Vines and epiphytes are common, the latter including ferns and many orchids. New Zealand shares only a few genera with these forests, for example, *Agathis, Elaeocarpus, Dysoxylum* and *Beilschmiedia*.

### Montane Zone

The montane zone ranges from about 700 to 3200 m altitude. The lower montane forests reach up to an average altitude of 1700 m, are 20–25 m in height and are generally without emergents. Palms and tree ferns are uncommon.

Some of the tree genera shared with New Zealand are *Elaeocarpus, Litsea, Weinmannia, Pittosporum* and *Schefflera*. Shared vine genera are *Parsonsia* and *Clematis*.

In places dense stands of *Araucaria cunninghamii* and *A. hunsteinii* are emergent over lower montane forest species. The araucarias establish in forest openings, but eventually give way to broadleaves in the course of forest succession.

'Oak forests', which are also found in the lower montane zone, generally on ridge crests, are dominated by relatively small-leaved, evergreen species of the oak-related genera *Castanopsis* and *Lithocarpus*. Oak forests are usually fairly open with few epiphytes and lianes. The rather sparse subcanopy includes tree ferns, *Phyllocladus hypophyllus* and species of *Rhododendron* and *Vaccinium*.

Mid montane forests, from about 1700 m to 2800 m, are moister than those at lower montane levels and have an abundance of terrestrial and epiphytic ferns, mosses and lichens. The commonest forest type at this level can be termed conifer broadleaf as conifers are frequent. They include species of *Podocarpus, Dacrycarpus, Papuacedrus* (related to *Libocedrus*) and *Phyllocladus*. Among broadleaf genera, species of *Weinmannia* are particularly frequent.

Species of *Nothofagus* of the '*N. brassii* group' are an important component in these forests, particularly on ridge crests where pure or nearly pure stands of *Nothofagus* occur.

We also recognise an upper montane conifer broadleaf forest in which liverworts and lichens are abundant on trunks and branches. Neither *Weinmannia* nor *Nothofagus* is present here.

**Subalpine Zone**

Unlike in New Zealand, *Nothofagus* does not extend up to the treeline. The subalpine zone from about 3200 to 3800 m is occupied by a low forest about 10 m high with emergents (mostly conifers) to 15 m. Again there is a mixture of conifers (*Dacrycarpus, Papuacedrus* and *Phyllocladus*) and broadleaves among which species of *Pittosporum, Myrsine, Schefflera, Rhododendron* and *Vaccinium* are prominent.

**Alpine Zone**

Above the subalpine forest and in frosty hollows within it are tussock grasslands similar in appearance to those of the New Zealand mountains. However, no species of the snowgrass genus (*Chionochloa*) are present. Instead the tussocks belong to such grass genera as *Poa, Deschampsia* and *Deyeuxia,* also represented in New Zealand and many other parts of the world. Scattered through the tussock grassland are several shrubs including species of *Coprosma, Gaultheria* and *Parahebe* and small trees of *Olearia spectabilis.* In places clumps of the fern *Papuapteris linearis* intermingle with the tussocks and at lower elevations a distinctive species of tree fern (*Cyathea atrox*) is scattered throughout. Above about 4000 m there is a short grassland with species of *Festuca* and *Poa* as well as a number of dwarf shrubs. In poorly drained places there are swamps and bogs: the swamps dominated by grasses and sedges; the bogs by cushion plants and small shrubs. The cushion plants include *Astelia papuana, Oreobolus pumilio* and *Carpha alpina.* The last species also occurs in Australia and New Zealand.

Above about 4200 m species of *Styphelia, Parahebe* and *Drapetes* form close mats on rocky sites with small grasses and species of *Ranunculus, Parahebe* and *Potentilla* on level areas and in hollows.

The alpine flora of New Guinea is not rich in species and has links with both north and south temperate regions. An unusually high proportion, 30 per cent, of the species are found outside New Guinea.

## New Caledonia and Other Pacific Islands

### New Caledonia[188]

Although only 400 km long, New Caledonia is composed of continental rocks similar to those of New Zealand and has a rich and, in many ways, remarkable flora. Conifers are particularly prominent with araucarias, tall and pencil-like or of candelabra form, providing a distinctive character to many landscapes. Indeed, of the 19 species of *Araucaria* recognised for the world 13 are restricted to New Caledonia. Several species of *Agathis* (kauri) are also present, several genera of the family Podocarpaceae, including the only known parasitic conifer (*Parasitaxus ustus*) and 3 species of *Libocedrus*. Several old southern families of flowering plants are also strongly represented — Myrtaceae, Proteaceae, Epacridaceae, Cunoniaceae. The primitive family Winteraceae is centred in New Caledonia with 4 genera and about 16 species.

For the most part New Caledonia has a quite steep topography, but the mountains up to 1600 m in altitude are not high enough to support alpine vegetation. On the drier west of the island and on burnt sites in the east there are extensive open woodlands of Australian aspect dominated by *Melaleuca quinquenervia* with its spongy fire-resistant bark. As in Australia, the *Melaleuca* also grows in swampy sites and it has been suggested that prior to the arrival of human beings it may have been largely restricted to such places, becoming more widespread following destruction of rain forests on better drained sites by fires of human origin.

At lower elevations closed forests which are still intact share many genera with similar forests in Australia and New Guinea and to a lesser degree New Zealand.[189] Strangling figs may occur as emergents. In places, particularly on ridge crests, there are also species of *Araucaria* and *Agathis*. A number of endemic palms are prominent as are some tree ferns, including *Cyathea novaecaledoniae* with its distinctive greyish trunks which attain heights of up to 30 m. The ferns *Asplenium nidus* and *Drynaria rigidula* are frequent as nest epiphytes.

At higher altitudes the montane rain forests are lower in stature, different in species composition and have more New Zealand links. The strangling figs of lower elevations give way to *Metrosideros* and the related *Carpolepis*, some species of which are frequently or occasionally initially epiphytic. Similarly the nest fern epiphytes give way to *Astelia novae-caledoniae*. *Weinmannia* and other members of the Cunoniaceae, tree ferns and conifers of the family Podocarpaceae are important components of these forests.

*Nothofagus* occurs in patches mixed with other genera or in pure ridge crest stands, although not on all mountains. The species of *Nothofagus*, as in New Guinea, belong to the *N. brassii* group and some are notable for having unusually large leaves particularly at the juvenile stage (Fig. 70).

The most notable physical feature of New Caledonia as far as the plants are concerned is the unusual extent of ultramafic or 'serpentine' rocks, which occupy a third of the island. Unlike some similar sites elsewhere in the world New Caledonian serpentine supports a rich and distinctive flora[190] in which conifers, *Nothofagus*, and the families Myrtaceae, Proteaceae, Epacridaceae and Casuarinaceae are prominent. Shrub and sedge associations are widespread, but there are also quite dense forests particularly at higher elevations.

### Fiji[4]

Further into the Pacific to the east the Fijian Islands are largely volcanic although, according to some geologists, continental rocks underlie them. A small continental element in the flora includes ten conifers — a species of *Agathis* and several podocarp species in several genera — and the primitive flowering tree *Degeneria vitiensis* placed in its own family close to the Winteraceae.

### Tahiti[4]

Much further to the east, entirely volcanic and remote from any continent, Tahiti is the highest and largest island of French Polynesia. Geologists believe that Tahiti, as well as other similar islands, rose above the sea in isolation. It would follow that the ancestors of its flora must have arrived by overseas dispersal from already existing islands near and far and directly from continents or indirectly via intervening islands. Possible involvement of a proposed former continent Pacifica was considered in Chapter 1.

Tahiti, along with other very isolated islands, completely lacks certain plant groups. It has no conifers, no species of *Nothofagus*, no members of the Winteraceae or related primitive families nor of the southern family Proteaceae. Of other important southern families the Myrtaceae is represented by *Metrosideros* and *Decaspermum,* the Cunoniaceae by *Weinmannia* and the Epacridaceae by *Styphelia.* Near sea level, native vegetation is virtually non-existent, but at higher levels there is a good cover. The forest is low, and in hollows and valleys is dominated by *Weinmannia parviflora* which, like the New Zealand *Weinmannia racemosa,* has compound juvenile and simple adult leaves and also often commences life as a low tree fern epiphyte. A long leaved *Freycinetia* is abundant and there are small orchid and fern epiphytes. Among small trees and shrubs are one large and one small-leaved species of *Myrsine,* a *Meryta* and in places a large-leaved *Coprosma* and the small tree *Fuchsia cyrtandroides.* In some valleys there are dense tree fern forests which may follow forest clearance as in New Zealand. The tree ferns are species of *Cyathea* and at least one has the unusual habit of forming more or less spherical bulbils below the leaf crown which drop off to grow into new plants.

On narrow ridge crests species of *Metrosideros* are conspicuous along with some of the plants from the valleys and in more open places a small-leaved *Styphelia.* One species each of *Ascarina, Macropiper* and *Ilex* are also present. On the ground and sometimes epiphytic is an *Astelia* and in moist places an *Elatostema.*

The Tahitian peaks are not high enough for the latitude to support alpine plants.

### Hawai‘i[3]

The Hawaiian Islands lie in the tropical north Pacific. The islands in the group are mostly larger than Tahiti and the largest and southernmost, Hawai‘i itself, has dome-like volcanoes, some still active, rising to over 4000 m. This is high enough to provide alpine conditions.

Hawai‘i shares with New Zealand and Tahiti such genera as *Metrosideros, Astelia* and *Coprosma,* but has more species of them than Tahiti. It lacks *Weinmannia,* a genus important in some Tahitian and New Zealand forests. Species and varieties of *Metrosideros* dominate in most Hawaiian forests, particularly on the island of Hawai‘i where some forms colonise lava flows and may form trees up to 30 m in height. In time other less light-demanding and shorter species enter the forests. They belong to such genera as *Sapindus, Ilex, Osmanthus* (some unite the Hawaiian species of this genus with *Nestegis* of New Zealand), *Pittosporum, Myrsine* and

*Elaeocarpus.* Tree ferns of the genus *Cibotium* may be common and sometimes bear a species of *Cheirodendron* (a genus related to *Pseudopanax*) as an epiphyte. There are also a few larger-leaved species of *Coprosma,* a long leaved *Freycinetia* and several species of *Astelia* both terrestrial and epiphytic. The lobelia family has undergone remarkable development in Hawai'i with several genera and many species of shrubs and small trees, some of which are found in moist forests.

Apart from *Metrosideros* the only other large tree species is *Acacia koa.* This may be a component of *Metrosideros* forests or may dominate more open woodlands.

Above tree limit on the high volcanoes of Hawai'i and Maui there is low shrubby vegetation on well drained sites reminiscent of the mountain shrubland of the volcanic Mt. Ruapehu of New Zealand. Close inspection however reveals very few species including a *Sophora,* a small-leaved *Styphelia,* a *Vaccinium* with bright red, shiny berries, a few small-leaved coprosmas, and a few ferns.

At higher levels still is a volcanic desert with few alpine species — a few grasses, small ferns, a few herbs in such genera as *Fragaria* and *Geranium,* and small shrubs such as the trailing small-leaved *Coprosma ernodioides.* The most remarkable alpine is the silversword (*Argyroxiphium sandwicense*) of the daisy family which is found only on Mt. Haleakela on Maui. It is somewhat like the celmisias in New Zealand with sword-like leaves, but much larger, with the silvery-white leaves almost spherically arranged. The tall reddish flower heads are not unlike those of some of the pleurophyllums of the New Zealand subantarctic.

Subalpine and alpine bogs have a much closer cover of plants. In less wet places there are dwarf shrubs of *Metrosideros* and other genera, but the wettest parts are essentially cushion bogs with cushions of *Oreobolus,* some grasses and species of *Astelia.* Scattered rosette herbs include droseras (sundews) and species of *Plantago, Geranium* and *Brachycome.*

## Temperate South America and Antarctica

### South America[191]

Temperate South America, much wider than New Zealand in the north but narrowing southwards, is in some respects like the South Island of New Zealand on a much larger scale. The Andes and the Southern Alps both lie more or less at right angles to the prevailing westerlies, so climates are wet in the west and much drier in the east. The Andes are much higher than the Southern Alps in the north at about 34°S, but they diminish in height southwards.

To the east of both mountain ranges are plains, but those in South America are much more extensive and in places drier than those of New Zealand. To the west in the South Island, and at equivalent latitudes in South America, the mountains come close to the sea with at most a narrow coastal plain in both places. Further to the north in South America the pattern is somewhat different as here there is a central valley running north-south, which is enclosed by the Andes to the east and a coastal range to the west. The major difference between the two regions is that in South America there is continuous land from 47 to about 55°S, whereas these latitudes in the New Zealand region are largely occupied by sea with only scattered islands.

The capital of Chile, Santiago, and the northern tip of New Zealand lie at the same latitude, 34°S, but there is little correspondence between the plants of the two places. The central valley of Chile at this latitude is semi-arid and going northwards the aridity increases until perhaps the driest deserts in the world are reached in Peru. Going southwards, climates become moister and from about 40°S they are comparable in their range of temperature and moisture regimes to those of New Zealand in the same latitudes.

On the seaward slopes of the coastal ranges from about 40 to 47°S and ranging in altitude from 300 to 500 m there is a moist closed forest without *Nothofagus*. Similar forests are to be found on the eastern flanks of the coastal range and the western slope of the Andes, although in these cases the deciduous *Nothofagus obliqua* is present as an emergent. This forest type is comparable both vegetationally and floristically to the conifer broadleaf forest of New Zealand, although its conifers in fact are few and inconspicuous.

Compared with forests at similar latitudes in New Zealand there are fewer species overall in the Chilean forests as well as fewer vines and epiphytes. Shared tree and shrub genera are *Podocarpus, Weinmannia, Laurelia, Pseudopanax* and *Griselinia*. Shared epiphyte genera are *Griselinia, Luzuriaga* and the ferns *Asplenium, Grammitis* and *Hymenophyllum*. Chilean vines are not related to those of New Zealand. A sight which is strange to New Zealand eyes is the presence in this South American forest type of clumps of the bamboo *Chusquea quila* and the substitution for our asteliad nest epiphytes of the bromeliad *Fascicularia bicolor* and for our ground astelias another bromeliad *Greigia sphacelata*. These bromeliads can be considered southern outliers of a family of specialised epiphytes that is richly developed in the Amazon rain forests.

*Nothofagus* forests are also extensively developed in southern South America.[192] There are 10 species of *Nothofagus* altogether, 2 belonging to the *N. menziesii* group and 8 to the *N. fusca* group. Seven species are deciduous (both of the *N. menziesii* group species and 5 of the *N. fusca* group), generally in response to seasonal drought or cold.

Above the lowland forest on the seaward slopes of the coastal range there is first a zone of evergreen *Nothofagus* involving two species in succession, then below treeline the deciduous *N. antarctica*. This is similar to the *Nothofagus cunninghamii/N. gunnii* pattern in Tasmania. On the inner coast range and the western Andes the pattern is different. Above the lowland forest the zonation sequence is deciduous, evergreen, then deciduous *Nothofagus*.

The two deciduous species of the *N. menziesii* group, unlike *N. menziesii* itself, are tolerant of some seasonal drought and used to form forests in the central valley from 38 to 41°S and further north at higher altitudes on the coastal range. *Sophora microphylla* and species of *Aristotelia* and *Muehlenbeckia* are sometimes associated with these species of *Nothofagus*.

Conifers associated with *Nothofagus* forests are *Podocarpus* species in the subcanopy, *Saxegothea conspicua* in the canopy and *Araucaria araucana* as an emergent near treeline in *Nothofagus pumilio* forests.

Going southwards in South America beyond New Zealand latitudes *Nothofagus* forest continues as the only type of forest to 55°S or about the latitude of Macquarie Island. Only three species of *Nothofagus* are present in these latitudes — the evergreen *N. betuloides* and the deciduous *N. pumilio* and *N. antarctica*.

From about 52 to 55°S the sequence from west to east is: an open moorland with little forest; evergreen *N. betuloides*; deciduous *N. pumilio* and *N. antarctica*; and short tussock grassland on the eastern plains. It has been suggested that the lack of forest in the west is due to the cloudy and thus cooler climate.[192] As conditions become drier and somewhat warmer to the east, evergreen *Nothofagus* forest can be supported, then deciduous *Nothofagus* forest and finally in the driest conditions of all only grass and shrubland. Alternatively, it has also been proposed that the lack of forest in the east is due to the diorite rocks there which form a thin infertile soil.[191]

The moorlands are of particular interest as they have cushion bogs very similar ecologically to those of New Zealand and Tasmania and with related floras. Genera shared with New Zealand are *Oreobolus, Gaimardia, Phyllachne, Donatia, Astelia, Drosera* and *Lepidothamnus. Lepidothamnus fonckii* has a very similar growth habit to the pygmy pine (*L. laxifolius*) of New Zealand.

The short tussock grasslands of the eastern plains are similar in general appearance to those of the eastern South Island and also involve species of *Festuca*. The associated shrubs and herbs are mostly unrelated to those of New Zealand with the exception of small species of *Discaria* (matagouri in New Zealand).

Above treeline in southern South America are extensive alpine areas.[193] There does not seem to be a well developed mountain shrubland zone as in some New Zealand mountains, although reduced, shrubby forms of *Nothofagus antarctica* may play that role. In poorly drained places cushion bogs similar to those already described occur, and on rocky sites fellfield with a predominance of cushion plants including large specimens of *Azorella* and the related *Bolax*. In the north where the Andes are at their highest there are extensive screes with some specialised plants similar to those of New Zealand, but not related to them.[194]

### Antarctica[195]

The southern tip of South America and the tip of the Antarctic Peninsula (Grahamland) of Antarctica are only about 800 km apart and are connected by a loop-like submarine ridge bearing several islands. Unlike adjacent South America, however, the present day plants of Antarctica can be dismissed quite briefly. Moss cushions and lichens grow on exposed rock on the Antarctic Peninsula and some rocks free of ice on the main part of Antarctica, particularly in the strange 'Dry Valleys' near the Ross Sea. The only flowering plants are two small species on the Antarctic Peninsula: a grass, *Deschampsia antarctica*, and a chickweed relative, *Colobanthus quitensis*, both of which are also found at the southern end of South America and elsewhere.

CHAPTER 12

# Plant Fossils and the Story They Tell

In the preceding chapters we have considered the present day plants and plant communities of New Zealand in some detail and those of some other lands more briefly. Some readers may now wonder how the present botanical patterns developed over geological time and in particular how such widely separated places within and bordering the South Pacific came to have related plants in similar communities. This is a difficult question to answer, but with plant fossils we do have a sort of time machine, even though the information it provides is very incomplete and sometimes difficult to interpret. Trying to reconstruct the botany of the past with fossils is a bit like trying to put together large and complex jigsaw puzzles, when many of the parts are missing and some of those we have are difficult, perhaps impossible, to place. Nevertheless it is often possible to find and correctly fit a sufficient number of pieces to gain at least a general impression of the overall pattern and a very clear idea of some details.

## How Do Plants Become Fossils?

### Macrofossils

First of all it is usually detached bits of plants rather than whole plants that become fossils, and this of course makes identification more difficult. Of macrofossils (visible to the naked eye) the most common are leaves, which often detach cleanly from stems along a special layer of weak cells, then twigs, and, less commonly, cones of conifers and fruits and seeds of flowering plants. Unfortunately flowers, which are the most reliable means of identification, are mostly soft tissued and often decay before they can become fossils.

Fossilisation frequently takes place at sites in the lowlands where there is deposition of the products of the weathering and erosion of rocks — clay, silt, sand, and so on. Such sites are the estuaries and deltas of rivers, river flood plains, ponds and lakes. Plant parts fall, blow or are washed into these bodies of water, sink to the bottom and become buried under layers of sediment. As the sediments become deeper and deeper the lower layers are compressed by the weight of those above, they undergo physical and chemical changes and eventually become solid rock. Any fossils also become greatly compressed and so altered that often all that is left is a carbonaceous film with no cell structure at all. Such fossils are termed compressions (Fig. 121) and can be revealed by splitting sedimentary rocks along their bedding planes. Sometimes there is no original or even substituted plant material remaining, but there may still be an imprint of the plant part. Such fossil

*Figure 121 (left)* Fossil leaf compression, about 7x4 cm, of the extinct *Nothofagus oliveri,* from Nuggety Creek, near Murchison. Mid-Miocene age. Photo: J. E. Casey.

*Figure 122 (above)* Pollen of *Zygogynum baillonii* of New Caledonia. *Zygogynum* belongs to the primitive family Winteraceae and at the present day is restricted to New Caledonia. The distinctive fossil pollen of this or a related genus has been found in New Zealand, in a fossil assemblage of Middle Pliocene Age. Photo: F. B. Sampson.

imprints are known as impressions. Compressions and impressions, particularly of leaves, can reveal a great deal — shape, margin form and sometimes complete detail of vein systems. If the resistant waxy surface layers of leaves (known as cuticles) are preserved, the impressions on them of long gone epidermal cells can be studied and their distinctive shapes and arrangements determined. Sometimes with three dimensional plant parts, such as portions of trunks or branches, the specimen may rot away completely, while the surrounding sediment solidifies, leaving a cavity lined with an impression of its surface. If the cavity later becomes filled with sediment, this in time solidifies into a cast or replica of the original.

The slow but steady accumulation of organic material in swamps and bogs can also lead to the formation of fossils as the conditions at such sites are often acidic and inhibit decay organisms. The plant remains, which form peat below the living plants at the surface, are termed subfossils. If peat becomes deeply buried, particularly under layers of inorganic sediments, pressure and heat gradually convert it into coal and the plant parts steadily lose their structure until none remains in the highest grade coals.

Plants may also be fossilised in fine textured volcanic ash redistributed by the heavy rain often associated with eruptions.

Sometimes the complete cellular structure of fossils is preserved. This happens when plant material, most often in swamps, becomes impregnated and replaced by silicates (sometimes near silica springs), carbonates and similar compounds in solution. After such an event the replaced plant material becomes resistant to physical and chemical change and the preservation of cells and sometimes cell contents can be remarkably good even in fossils hundreds of millions of years old. By special techniques thin sections can be made of these well preserved fossils and cell structure can be studied in detail.

**Microfossils**

These are principally the spores of ferns, bryophytes and other groups and the pollen grains (Fig. 122) of conifers and flowering plants. They are useful in interpreting past floras because of their great resistance to decay (with some exceptions), and because of a number of distinctive and constant characteristics which often enable them to be identified to family, genus or, less often, species. The resistance to decay derives from a wax-like compound in the cell walls. The features useful in identification are the shape of the grain (spherical, triangular, bean-like etc); the number, form and arrangement of germination pores; the structure of the wall; and the sometimes intricate pattern of sculpturing on the surface of the spore or grain.

Some plants are more likely to be fossilised than others. Lowland plants are better represented than alpine plants because it is in the lowlands that deposition is prevalent. The fossilisation process may begin in suitable alpine sites such as bogs and ponds, but it is likely to be interrupted by erosion at an early stage. Firm or even hard textured plant parts make better fossils than delicate parts that tend to shred or decay rapidly. Plant parts which separate readily, such as leaves with abscission layers, are more common as fossils than parts which normally remain attached and slowly decay. Spores and pollen which are wind dispersed are produced in great quantities and spread for great distances, so are often over-represented in the fossil record by comparison with bird, water or insect-pollinated species.

## Lady Lake

An interesting study was recently carried out at the small Lady Lake in north Westland, New Zealand, in which the relative representation of potential macro-fossils[196] and micro-fossils[197] in the lake sediments was compared with the representation of the present day flora in the region. Surrounding the lake is kahikatea (*Dacrycarpus*) swamp forest, conifer broadleaf forest and swamp. With potential macro-fossils it was found that most trees and shrubs were represented. Exceptions were putaputaweta (*Carpodetus serratus*), *Fuchsia excorticata,* kaikomako (*Pennantia corymbosa*), wineberry (*Aristotelia serrata*) and *Coprosma tenuicaulis,* which all have leaves that decay readily. Kahikatea was under-represented and there seems no obvious explanation for this. Some trees and shrubs appeared to be better represented in the lake sediments than in the surrounding forest — ma̅tai (*Prumnopitys taxifolia*), Hall's totara (*Podocarpus hallii*), *Phyllocladus aspleniifolius,* manuka (*Leptospermum scoparium*), kanuka (*Kunzea ericoides*), southern rata (*Metrosideros umbellata*) and *Coprosma propinqua*. Forest floor herbs were not represented in the lake sediments although some high herbaceous epiphytes, *Lycopodium varium* and the orchids *Earina* and *Dendrobium* were. This may be because the latter are better situated to fall or to be blown into the lake.

Aquatic and swamp species were poorly represented in the lake litter. Many of them are soft-tissued and decay readily and in the cases of those that are very fibrous, such as *Phormium tenax,* dead leaves remain attached to the plants and gradually decay.

Some of the litter would be brought into the lake by streams, especially when in flood, some by surface wash during heavy rain, some by direct fall from overhanging plants and some by wind, particularly during gales.

Spores and pollen were sampled from the surface of the lake sediments and from moss cushions. They gave a picture broadly similar to that of the macro-plant remains, but there were some differences. By contrast with the lake litter, *Quintinia, Cyathea smithii* and kamahi (*Weinmannia*) were under-represented. The epiphytic orchids were not recorded at all. The two types of remains were in agreement in under-representing *Dacrycarpus* and sedges. It has been suggested that at least some of the under-representation of these species may be due to the rapid decay of their spores or pollen under lake conditions. Some of these species are well represented in peat where the acid conditions would aid their preservation.

## The Fossil Record of New Zealand and Other Southern Lands

Despite problems of under, over or lack of representation, difficulties in identification due to incomplete and sometimes imperfect preservation, or to some of the fossils being of extinct plants with no modern relatives, it is possible to derive a general picture of the sequence of past botanical events in New Zealand and some related floras. Before we attempt to do this, however, it is necessary to review the ways in which the positions of crustal blocks in the South Pacific have changed over time,[198,199] as this is critical to the interpretation of the botanical trends.

### Crustal Movements

The modern floras are dominated by flowering plants with conifer and fern components, so we can begin our journey from the past at the time when flowering plants first began to achieve prominence — the middle of the Cretaceous period (Fig. 123) about 100 million years ago.

By this time Africa and India had long since separated from the great southern continent, Gondwana, but Australia, Antarctica and the New Zealand-New Caledonia crustal complex (Tasmantis) were still united (Fig. 3). The southern end of South America was probably joined with Antarctica at this stage and remained so until the Oligocene, 30 million years ago.

During the middle Cretaceous, the Australasian portion of Gondwana lay in higher southern latitudes with the northern parts of New Zealand and Australia at about 60°S and 45°S respectively.

In the late Cretaceous (80 million years ago) the New Zealand crustal block began to separate from Australia and Antarctica, moving like a hinge with the northern end of the Lord Howe Rise remaining adjacent to Australia. By the middle Paleocene (60 million years ago) northern New Zealand had reached 55°S and the Norfolk Ridge had separated from the Lord Howe Rise. It is possible that there were dry land connections at this time with Queensland, via the Lord Howe Rise, and New Caledonia via the Norfolk Ridge.

In the late Paleocene at about 50 million years ago Australia and Antarctica were moving apart with north Australia reaching 30°S. Northern New Zealand had attained 40°S by this time. One result of these crustal movements was the formation of the Lau and Tonga volcanic ridges near to and beyond the northern Norfolk Ridge. Fiji lay at the northern end of these ridges.

By the late Eocene at 38 million years ago Australia had not changed much in position. New Zealand had moved further north to approach its present northernmost position of 35°S and the Lau and Tonga ridges had moved eastwards away from the Norfolk Ridge.

The late Oligocene (29 million years ago) saw north Australia reaching 25°S and the activation of the Alpine Fault in the New Zealand region with movement southwards along it of most of the present South Island and the Campbell Plateau and Chatham Rise. By this time, most of the New Zealand-New Caledonia crustal block was submerged.

By the Late Miocene (10 million years ago) north Australia was at 15°S and the rearrangement of the New Zealand region along the alpine fault continued. Between this time and the present the current configuration was attained with north Australia reaching 10°S and New Guinea almost reaching the equator. In the last few million years crustal movements have raised the New Zealand and New Guinea mountains.

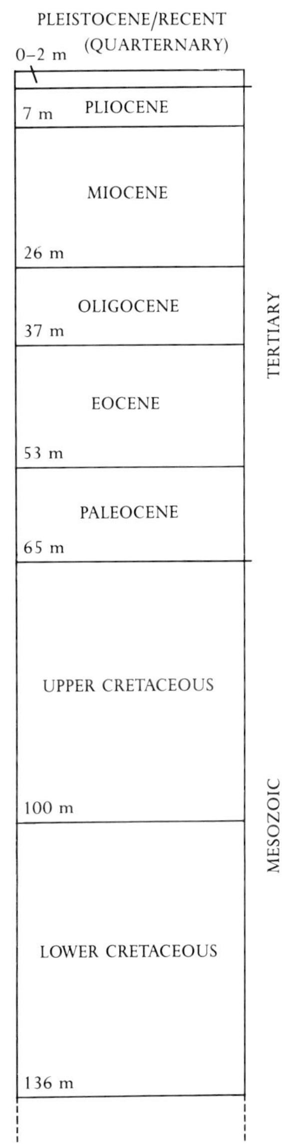


*Figure 123* Geological time scale.

## Botanical History

### Cretaceous

The fossil floras of New Guinea, New Caledonia and other smaller islands are too fragmentary and/or poorly known to provide any coherent picture. Australia, New Zealand, Antarctica and southern South America will be considered together for the Cretaceous period,[15,195,200] when they were still joined and shared a largely common flora. For the periods of the Tertiary era, when they had separated and their floras had undergone different trends, they will be treated individually.

In the middle Cretaceous when Australasia, Antarctica and southern South America were joined in high southern latitudes their climates appear to have been much warmer than those of similar latitudes today. The linked lands also seem to have been generally moist, even in parts now arid, and they shared a type of forest dominated by conifers and ferns with an increasing flowering plant component. The tree fern family Cyatheaceae was prominent and among conifers the Araucariaceae[201] were represented and the Podocarpaceae were becoming dominant. The earliest flowering plants are known mostly from pollen that cannot be matched with any present day types.

In the latest Cretaceous, when this part of Gondwana began to break up, flowering plants increased in the southern lands and a number of the micro-fossils and macro-fossils can be assigned to modern families, or in some cases genera. *Nothofagus* is a notable case with *brassii* group pollen appearing at about the same time in Australia, New Zealand and Antarctica (from fossil pollen near McMurdo Sound) and somewhat later in South America where it was associated with the *fusca* and *menziesii* groups. Pollen from these two groups appears in New Zealand at the end of the Cretaceous and in Australia in the early Tertiary. Pollen assignable to the Proteaceae is also found in the four regions in the Late Cretaceous. In South America, New Zealand, and Australia in particular, a number of other angiosperm fossils of late Cretaceous age have been found but these have not yet been identified. Some of the families that have been recognised, however, include the Winteraceae, Epacridaceae, Chloranthaceae (*Ascarina*), Loranthaceae (mistletoes), (?)Scrophulariaceae, Lauraceae, and, rather surprisingly, several families which are largely herbaceous at present, although they may not have been so then

— Ranunculaceae, Haloragaceae (*Gunnera*), Cruciferae and Caryophyllaceae.[15]

**Tertiary Australia[198,202,203]**

For most of the Paleocene Australia was still joined to Antarctica although a rift valley was developing between them. Southern Australia was at 65°S but there is evidence that sea temperatures were of subtropical warmth so temperatures on land were probably warm also. Microfossils indicate widespread forests dominated by Podocarpaceae; *Nothofagus* was rare. Araucariaceae, Proteaceae, and Myrtaceae were also represented and in addition *Anacolosa* and possibly *Cupania,* genera now restricted to the tropics. Forests of this type were also present in Central Australia so the general climate was moist as well as warm. High rainfall, warm temperatures and low relief at this and later times resulted in many places in intensely leached infertile soils which have persisted to the present day.

At the beginning of the Eocene, final separation of Australia and Antarctica began. At first conditions continued to be warm and moist and further tropical genera were added to the forests such as *Bombax* (the kapoc genus), and the tropical mangrove palm *Nipa.* Conifers and *Nothofagus* were still poorly represented. From the middle Eocene the climate cooled, a number of the tropical genera disappeared from southern Australia, and pollen of *N. brassii* group increased in importance.

During the Oligocene the climate cooled further and the sparse fossil record indicates similar floras to the late Eocene although with some reduction in diversity. It is suggested that with the lower temperatures the climates would also have been drier and perhaps relatively arid in northern parts. An ice cap probably began to develop in Antarctica at this time.[198]

In the early Miocene the climate became distinctly warmer and moister than it was in the Oligocene. The fossils suggest extensive moist forests over southern Australia with *Nothofagus* of the brassii group, conifers, Myrtaceae and Lauraceae prominent. In central Australia, localised forests of the same type occurred but so did open habitats as indicated by the record of *Acacia, Casuarina* and grasses. Temperatures dropped again by the late Miocene when Australia had reached its present latitudes and the Antarctic ice sheet had reached its present dimensions. The rather limited evidence[198] suggests a retreat of rain forest with the cooler and drier climates.

The Pliocene also began with a warmer phase resulting in some rain forest expansion, but eventually the temperature declined leading to the severe glacial/interglacial oscillations of the Pleistocene. This finally led to the widespread disappearance of rain forest and expansion of plants adapted to arid conditions. These included *Eucalyptus, Casuarina, Acacia* and various Proteaceae. Most of the arid and semi-arid floral components in Australia are thought to be derived from families of early rain forest origin adapted to heavily leached infertile soils during the Tertiary.

**Tertiary Antarctica[195]**

A site near McMurdo Sound (presently at 78°S) yielded a Cretaceous flora of Podocarpaceae, *Nothofagus* and Proteaceae. Palm pollen is recorded in early Tertiary strata from the same locality, indicating a relatively mild climate, even though the latitude was much the same as the present day. The richest fossil floras of Antarctica are from Seymour Island, off the Antarctic Peninsula at 64°S, where leaves and pollen have been recovered. The date of these deposits is not certain, but it seems likely to be lower Tertiary. The flora is reasonably diverse and of rain forest character, although 23 species cannot be identified and of the remainder 27 are ferns including Cyatheaceae and Schizaeaceae. There is a strong conifer component including the genera *Araucaria, Agathis, Dacrydium, Phyllocladus* and *Podocarpus*. Angiosperms include *Nothofagus* of the *fusca* and *brassii* groups and the following families: Cunoniaceae, Lauraceae, Monimiaceae (*Laurelia*), Myrtaceae, Proteaceae (*Knightia*), Winteraceae, Loranthaceae, Leguminosae, Aquifoliaceae (includes *Ilex*, the holly genus), Cruciferae and Cyperaceae (sedges). This assemblage is similar to that produced by the rain forests of Australia in the early Tertiary with the difference that there are no 'tropical' taxa, such as *Anacolosa* and *Cupania*.

The most recent record of vegetation in Antarctica comes from pollen in Ross Sea deposits of the late Oligocene age. At this stage an ice sheet was beginning to form. The assemblages were dominated by *Nothofagus* pollen (with the *fusca* group most common) plus lesser amounts of Proteaceae, Myrtaceae and Podocarpaceae.

## Tertiary South America[195,198]

The general progression in South America seems to have been similar to that of Australia although arid climates, while strongly developed in places, did not become so widespread as in Australia.

In the Eocene, genera and families now largely restricted to tropical northern South America appeared in the fossil records of Chile and Argentina. This is comparable to the appearance of a tropical element in the southern Australian floras at about the same time. Similarly, with cooling from the Oligocene on, the tropical element disappeared from southern South America and southern rain forest groups — Podocarpaceae, *Nothofagus,* Proteaceae, Cunoniaceae — became increasingly important.

## Tertiary New Zealand[15,204,205,206]

Mountains were raised in the New Zealand crustal complex (Rangitata orogeny) in the late Jurassic and early Cretaceous when it was still part of Gondwana. By the late Cretaceous, when separation of New Zealand from Gondwana began, it is believed the mountains had been largely eroded, so the physical background to most of the Tertiary fossil record we are about to consider is one of a low lying peneplain with strongly oceanic climates.

During the Paleocene, conifers of the family Podocarpaceae were dominant and *Nothofagus* of all three groups was fairly rare. Proteaceae were quite common and, as in Australia, *Anacolosa* appeared as well as a number of palms including probable *Nipa*. *Casuarina* also appeared and remained a significant element for most of the Tertiary. *Casuarina* no longer occurs in New Zealand as a native genus, but fossil pollen considered to be *Casuarina* had been known for some time. The recent discovery of impressions of the distinctive cone-like fruits of *Casuarina* and the related genus *Gymnostoma* has confirmed the presence of the family in New Zealand.[207] The Myrtaceae also appear for the first time in the Paleocene with pollen referable to both *Leptospermum* and *Metrosideros* having been identified.

In the Eocene the podocarps declined in importance, although *Phyllocladus* is recorded for the first time. *Nothofagus* became dominant, firstly the *fusca* group and latterly the *brassii* group. *Cupania* and *Bombax* appeared, rather later than in Australia, along with a number of new proteaceous genera and *Freycinetia, Astelia, Quintinia,* possibly *Phormium,* and the family Araliaceae (includes *Pseudopanax*). A genus now extinct in New Zealand, *Ilex* (which includes the holly), is also first recorded in the Eocene. Cooling at the end of the Eocene may have been the reason for the extinction of *Anacolosa, Nipa* and a number of Proteaceae.

During the Oligocene, New Zealand reached its present latitudes and also suffered its greatest reduction in area as a result of sea encroachment. The cooler climate of the late Eocene seems to have continued into the Oligocene, perhaps partly due in New Zealand's case to the development of cooler westerly winds and currents following the separation of Australia and Antarctica. The forests were dominated by the *Nothofagus brassi* group, although the *N. fusca* group, *Casuarina,* Myrtaceae, Palmae and Podocarpaceae were also prominent. Among genera to first appear in this period are *Weinmannia, Elaeocarpus, Myrsine, Fuchsia, Coprosma, Laurelia* and *Epilobium*.

The Compositae (daisy family) made its first appearance in the late Oligocene.

Warmer climates prevailed in the early Miocene and the land began to rise. Swampy areas were common and surrounding them were dense floristically rich forests. Many of the species cannot be identified with modern plants, but several *Nothofagus brassii* group species were common together with Myrtaceae (possibly including *Eucalyptus*), *Casuarina,* Podocarpaceae and tree ferns. Palms also were common and included a small fruited species of coconut in the northern North Island — *Cocos zeylanica* (Fig. 124). Among genera recorded for the first time are: *Cordyline, Ripogonum, Dysoxylum, Alectryon, Macropiper, Pittosporum, Muehlenbeckia and Melicytus*. An interesting incomer is *Acacia,* no longer present in New Zealand.

Temperatures declined towards the end of the Miocene and this trend continued into the Pliocene. Uplift continued during the Pliocene and the Southern Alps and other ranges were formed. This phase of uplift is known as the Kaikoura orogeny and continues today. Extensive development of alpine vegetation, as we have discussed earlier, would have commenced from this time. In the lowlands the *Nothofagus brassii* group was prominent in the early Pliocene but only in the northern North Island. South of there Myrtaceae, rimu (*Dacrydium cupressinum*), *Nothofagus fusca* group and ferns were common. By the end of the Pliocene a number of forest genera and species had become extinct in-

*Figure 124* Two fossil coconuts of the extinct *Cocos zeylanica*. The larger of the two nuts is only about 4 cm across. The fossils are associated with a coal seam of Miocene Age at Cooper's Beach in the far north of the North Island. The coal seam dips below the sea and the small coconuts are washed onto the beach from time to time.
Photo: J. E. Casey.

cluding most of the *Nothofagus brassii* group, *Bombax, Cupania* and many palms.

During the Pleistocene the climate deteriorated greatly with a succession of long glacials and shorter interglacials. The last species of the *Nothofagus brassii* group disappeared together with *Microstrobus* and *Microcachrys* (now restricted to Australia) of the Podocarpaceae, *Casuarina, (?)Eucalyptus, Acacia* and all Proteaceae except two species, namely *Knightia excelsa* and *Toronia toru*. With each glacial the alpine vegetation expanded and diversified; with each interglacial the forests expanded from refugia (albeit reduced in diversity) to largely reclothe the landscape.

### Disjunct Distribution Patterns

Certain present day anomalies in the distribution patterns of New Zealand species are thought by some to have been caused by these climatic fluctuations.

Wardle[108] noted that the northern North Island and northern and southern South Island regions have much higher numbers of endemic species (90–110) than the southern North Island and central South Island (25–30). The endemics of the northern North Island are mostly woody and lowland and those of the northern and southern South Island mostly herbaceous and alpine.

Wardle also observed that there are a number of species disjunct between the regions of high endemism, for example tanekaha (*Phyllocladus trichomanoides*) and *Astelia trinervia* between the northern regions of the two islands, and *Celmisia traversii* and species of *Nothofagus* between the northern and southern South Island.

It is suggested that these anomalies have the same explanation. During the last glaciation the central South Island with its high mountains would have been more severely glaciated than the neighbouring regions. Its flora would have been more reduced with the local extinction of some species that survived to the north and south. The low endemism of the southern North Island is more difficult to explain, but a somewhat lower treeline at the present time than would be expected and some fossil evidence indicates that climatic conditions were relatively severe there during the last glaciation.

Recently McGlone[208] has proposed a quite different explanation. He believes that the causes of these anomalous patterns are tectonic not climatic and must be sought much further back in time in the late Tertiary before the glacial/interglacial fluctuations had commenced. At this time much of the southern North Island had become submerged beneath the sea and this would explain the low endemism and some of the disjunctions. In the South Island the uplift of particularly high mountains in the central region would have greatly altered habitats there and 'as a result the once continuous, upland flora was split into northern and southern regions, where the rate of uplift was low and average age of surfaces older, and a rapidly uplifting central zone'.

## In Conclusion

The fate of the original vegetation cover of the various regions of the Australasian/Antarctic portion of Gondwana has depended on crustal movements and climatic change.

In the case of Antarctica, deterioration of the climate in the later Tertiary eventually led to the extinction of vascular plants.

The northward movement of Australia during the Tertiary counterbalanced the temperature decline to some degree, but the pronounced cold spells of the Pleistocene played some part in the disappearance of closed forest from most of the continent. However the major cause of the drastic vegetational changes from the late Tertiary would have been increasing aridity associated with declining temperatures. In the southern hemisphere, aridity is most marked along the Tropic of Capricorn which Australia now straddles.

The extent to which the rich rain forests of New Guinea correspond to those which were formerly widespread in Australia is uncertain. Until recently it was thought that much of New Guinea's rain forest flora was derived from south-east Asia after contact was made with that region. This was also thought to be the case with the rain forests of north Queensland, but this has recently been questioned as many of the genera concerned have a fossil record in Australia extending back to times when Australia and its northern portion which became New Guinea was still far south of Asia.[209] Certainly the rain forests of New Guinea in near equatorial latitudes would not have been greatly affected by cooler, drier climates during the Pleistocene. The alpine flora of New Guinea, as already mentioned, although interesting is not particularly rich and includes Asian and Australasian components.

Botanically speaking, New Caledonia is one of the most interesting fragments of Gondwana. Being tropical like New Guinea, although not equatorial, with a strongly oceanic climate due to its narrowness, it too has probably not been greatly affected by coolness or drought. As a consequence its inheritance of Gondwana conifer and angiosperm families would have survived better than those of Australia and New Zealand. This makes it of special importance to botanists in the latter countries attempting to reconstruct the history of their own floras, as now extinct genera in Australia or New Zealand may still have extant species in New Caledonia. A special development in New Caledonia has been the evolution of many species, mostly of Gondwana genera, tolerant of the soils of the extensive areas of ultramafic rock.

Finally we come to New Zealand. Like Australia, New Zealand's northward movement during the early Tertiary would have moderated the effects of any temperature declines and its oceanic climate would have countered any tendency to aridity. However, New Zealand had attained its present position some time before the more drastic climate deterioration of the Pleistocene; and the cold climates deriving from the combined effects of high mountains and glaciations, although ameliorated to some extent by an oceanic climate, greatly affected the vegetation. Forests retreated and some of their genera and species became extinct, while new species evolved in colder open habitats contributing to the present relatively rich alpine flora.

The native forests include genera with a long history in the region; the plants of the mountains are a more recent burgeoning of specialised forms in response to the uplift of the mountains and episodes of climatic refrigeration.

The recent advent of human habitation has greatly reduced New Zealand's natural cover — Polynesians achieved it with fire and, to more effect in a shorter time, Europeans with fire, the plough and the introduction of alien plants and animals.

The time has surely come to call a halt, to preserve those areas of largely intact vegetation we still have and to actively encourage the regeneration of native plants in areas abandoned from farming where fire and browsing have taken their toll.

# General References

Allan, H. H. *Flora of New Zealand,* Vol. 1. Government Printer, Wellington, 1961.

Cockayne, L. *The Vegetation of New Zealand,* Ed. 2. Engelmann, Leipzig, 1928.

Cockayne, L. *New Zealand Plants and their Story,* Ed. 4, (Ed. E. J. Godley). Government Printer, Wellington, 1967.

Cockayne, L. and Phillips Turner, E. *The Trees of New Zealand,* Ed. 5. Government Printer, Wellington, 1965.

Cooper, Dorothy. *A Field Guide to New Zealand Native Orchids.* Price Milburn. Wellington, 1981.

Crookes, Marguerite and Dobbie, H. B. *New Zealand Ferns,* Ed. 6. Whitcombe and Tombs, Christchurch, 1963.

Eagle, Audrey. *Eagle's Trees and Shrubs of New Zealand.* Collins, Auckland, 1975. Second Series, 1982.

Given, David R. *Rare and Endangered Plants of New Zealand.* A. H. & A. W. Reed, Wellington, 1981.

Heath, Eric and Chinnock, R. J. *Ferns and Fern Allies of New Zealand.* A. H. and A. W. Reed, Wellington, 1974.

Johns, John and Molloy, Brian. *Native Orchids of New Zealand.* A. H. & A. W. Reed, Wellington, 1983.

Laing, R. M. and Blackwell, E. M. *Plants of New Zealand,* Ed. 7 (Ed. E. J. Godley). Whitcombe and Tombs, Christchurch, 1964.

Mark, A. F. and Adams, Nancy. *New Zealand Alpine Plants.* A. H and A. W. Reed, Wellington, 1979.

Moore, Lucy B. and Edgar, Elizabeth. *Flora of New Zealand,* Vol. 2. Government Printer, Wellington, 1970.

Moore, L. B. and Irwin, J. B. *The Oxford Book of New Zealand Plants.* Oxford University Press, Wellington, 1978.

Philipson, W. R. and Hearn, D. *Rock Garden Plants of the Southern Alps.* Caxton Press, Christchurch, 1962.

Poole, A. L. and Adams, N. M. *Trees and Shrubs of New Zealand,* 4th Impression. Government Printer, Wellington 1979.

Salmon, J. T. *Collins Guide to the Alpine Plants of New Zealand,* Rev. Ed. Collins, Auckland, 1985.

Salmon, J. T. *The Native Trees of New Zealand.* A. H. & A. W. Reed, Wellington, 1980.

Sampson, F. B. *Early New Zealand Botanical Art.* Reed Methuen, Auckland, 1985.

# Notes and References

**Chapter 1**

1. Druce, A. P. 1984. 'Indigenous higher plants of N. W. Nelson and indigenous higher plants of New Zealand not in N. W. Nelson.' Circulated but unpublished checklist. Botany Division, D. S. I. R.
2. Broadleaf is a term applied to flowering plants (angiosperms), whose leaves are usually much wider than those of conifers.
3. Carlquist, S. 1970. *Hawaii: A Natural History.* Natural History Press, New York.
4. Carlquist, S. 1965. *Island Life.* Natural History Press, New York.
5. Ehrendorfer, F. 1979. 'Reproductive biology in island plants.' Pp. 293-306 in *Plants and Islands,* D. Bramwell (Ed.). Academic Press.
6. Lloyd, David G. 1985. 'Progress in understanding the natural history of New Zealand Plants.' *N. Z. Journ. Bot. 23*: 707–22.
7. Mabberley, D. J. 1979. 'Pachycaul plants and islands.' Pp. 259-277 in *Plants and Islands,* D. Bramwell (Ed.). Academic Press.
8. Gillett, G. W. 1972. 'The role of hybridization in the evolution of the Hawaiian flora.' Pp. 205-19 in *Taxonomy, Phytogeography and Evolution,* D. H. Valentine (Ed.). Academic Press.
9. Rattenbury, J. A. 1962. 'Cyclic hybridisation as survival mechanism in the New Zealand forest flora.' *Evolution 16*: 348-63.
10. Stevens, G. R. 1980. *New Zealand Adrift.* Reed, Wellington.
11. Godley, E. J. 1979. 'Flower Biology in New Zealand.' *N. Z. Journ. Bot. 17:* 441-66.
12. Bawa, Kamaljit S. 1979. 'Breeding systems of trees in a tropical wet forest.' *N. Z. Journ. Bot. 17:* 521–4.
13. Connor, H. E. 1985. 'Biosystematics of higher plants in New Zealand 1965-1984.' *N. Z. Journ. Bot. 23:* 613-44.
14. Primack, Richard B. 1983. 'Insect pollination in the New Zealand mountain flora.' *N. Z. Journ. Bot. 21:* 317-33.
15. Mildenhall, D. C. 1980. 'New Zealand late Cretaceous and Cenozoic plant biogeography: a contribution.' *Palaeogeography, Palaeoclimatology, Palaeoecology 31:* 197-233.

16. Craw, R. C. 1982. 'Phylogenetics, Areas, Geology and the Biogeography of Croizat: A Radical View.' *Systematic Zoology 31*: 304–16.
17. Melville, R. 1981. 'Vicarious plant distributions and paleogeography of the Pacific region.' Pp. 238-74 in *Vicariance biogeography: a Critique,* G. Nelson and D. E. Rosen (Eds.). Columbia University Press, New York.
18. Carey, S. Warren. 1983. 'The necessity for earth expansion.' Pp. 375-93 in *Continental Drift — a symposium,* Carey S. W. (Ed.). University of Tasmania.

**Chapter 2**

19. Guthrie-Smith, 1926. *Tutira. The Story of a New Zealand Sheep Station* (Ed. 2). Blackwood, Edinburgh.
20. Brown, G. S. 1960. 'A forester in New Zealand — First impressions.' *Malayan Forester* 23: 47-56.
21. Richards, P. W. 1952. *The Tropical Rain Forest. An Ecological Study.* Cambridge University Press.
22. Dawson, J. W. and Sneddon, B. V. 1969. 'The New Zealand lowland rain forest. A comparison with tropical rain forest.' *Pacific Science:* 131-47.
23. Dawson, J. W. 1986. 'Floristic Relationships of lowland rainforest phanerogams of New Zealand.' *Telopea 2:* 681-95.
24. Godley, E. J. 1985. 'Paths to maturity.' *N. Z. Journ. Bot. 23:* 687-706.
25. Philipson, W. R. 1964. 'Habit in relation to age in New Zealand trees.' *Maheshwari commemorative volume, Journ. Ind. Bot. Soc. 42A:* 167-79.
26. Friedmann, F. and Cadet, Th. 1976. 'Observations sur l'hetérophyllie dans les Iles Mascareignes.' *Adansonia, ser. 2, 15 (4):* 423-40.
27. Laing, R. M. and Blackwell, E. W. 1906. *Plants of New Zealand.* Whitcombe and Tombs, (Ed. 7, 1964).
28. Sampson, F. B. and McLean, Jean. 1965. 'A note on the occurrence of domatia on the underside of leaves in New Zealand plants.' *N. Z. Journ. Bot., 3:* 104-12.
29. Bussell, W. T. 1968. 'The growth of some New Zealand trees. 1. Growth in natural conditions.' *N. Z. Journ. Bot. 6:* 63-75.
30. Bussell, W. T. 1968. 'The growth of some New Zealand trees. 2. Effects of photoperiod and temperature.' *N. Z. Journ. Bot. 6:* 76-85.
31. Russel, R. S. 1936. 'The mechanism of leaf fall in certain New Zealand trees.' *Trans. Roy. Soc. N. Z. 65:* 407-21.

**Chapter 3**

32. Harrison-Smith, J. L. 1938. 'The kauri as a host tree.' *N. Z. Journ. For. 4:* 173-177.
33. Rhizomes are horizontal stems on or below the surface of a substrate, usually the ground.
34. Sporangia are organs containing spores.
35. Holloway, J. E. 1923. 'Studies in the New Zealand Hymenophyllaceae.' *Trans. N. Z. Inst. 54:* 577-618.
36. Bird, J. W. 1916. 'Observations on the lianes of the ancient forest of the Canterbury Plains of New Zealand.' *Trans. N. Z. Inst. 48:* 315-53.
37. Dawson, J. W. 1967. 'A growth habit comparison of *Metrosideros* and *Ficus*.' *Tuatara 15:* 16-24.
38. MacMillan, B. H. 1973. 'Biological flora of New Zealand 7: *Ripogonum scandens*. Supplejack, Kareao.' *N. Z. Journ. Bot. 10:* 641-72.
39. This is a curious phenomenon also to be found in the equally variable leaves of juvenile pokaka (*Elaeocarpus hookerianus*), juvenile *Pittosporum obcordatum,* seedling lancewood (*Pseudopanax crassifolius*) and others.
40. Eagle, Audrey. 1982. *Eagle's Trees and Shrubs of New Zealand.* Second Series. Collins.
41. Oliver, W. R. B. 1930. 'New Zealand epiphytes.' *Journ. Ecol. 18:* 1-50.
42. Hatch, E. D. 1948. 'The epiphytic orchids of New Zealand.' *Trans. Roy. Soc. N. Z. 78:* 101-5.
43. Belkin, J. N. 1968. 'Mosquito studies Vll. The Culicidae of New Zealand.' *Contr. Amer. Ento. Inst.* 3: 1-180.
44. Kirk, T. 1870. 'On the Botany of the Thames gold fields.' *Trans. N. Z. Inst. 2:* 89-100.
45. Dawson, J. W. 1966. 'Vegetative features of *Griselinia lucida*. A New Zealand shrub epiphyte.' *Tuatara 14:* 121-9.
46. Beddie, A. D. 1953. 'Root behaviour in *Metrosideros*.' *Wgtn. Bot. Soc. Bull. 26:* 2-6.
47. Druce, A. P. 1971. 'Uncle Rimus and Brer Ratas.' *Bull. Wgtn. Bot. Soc. 37:* 59-62.
48. Zotov, V. D. 1948. 'Rata the killer.' *Tuatara 1:* 36-8.
49. Dawson, J. W. 1985. '*Metrosideros bartletti* (Myrtaceae) a new species from North Cape, New Zealand.' *N. Z. Journ. Bot. 23:* 607-10.
50. Pope, A. 1924. 'The role of the tree fern in the New Zealand bush.' *N. Z. Journ. Sci. Tech. 7:* 52-61.
51. Page, C. N. and Brownsey, P. J. 1986. 'Tree fern skirts — A defence against climbers and large epiphytes.' *Journ. Ecol. 74:* 787-97.
52. Zahlbruckner, A. , Keissler, K. and Allan H. H. 1928. 'The epiphyllous lichens of Kitchener Park, Feilding, New Zealand.' *Trans. N. Z. Inst. 59:* 304-14.
53. Fineran, B. A. 1974. 'Parasitic flowering plants.' *New Zealand's Nature Heritage 23:* 637-41.
54. Moore, L. B. 1940. 'The structure and life history of the root parasite *Dactylanthus taylori*. *N. Z. Journ. Sci. Tech. 21B:* 206-24.
55. Philipson, W. R. 1959. 'Some observations on root parasitism in New Zealand.' *Trans. Roy. Soc. N. Z. 87:* 1-3.
56. Barlow, B. A. 1966. 'A revision of the Loranthaceae of Australia and New Zealand.' *Aust. Journ. Bot. 14:* 421-99.
57. Campbell, E. O. 1962. 'The mycorrhiza of *Gastrodia cunninghamii*.' Trans.; *Roy. Soc. N. Z. Bot. 1:* 289-96.
58. Campbell, E. O. 1968. 'An investigation of *Thismia rodwayi* and its associated fungus.' *Trans.; Roy. Soc. N. Z. Bot. 3:* 209-19.

59. Chandler, M. E. J. 1964. 'The Lower Tertiary Floras of Southern England. 4. A Summary and Survey of Findings.' London.

**Chapter 4**

60. Tanekaha is absent from the southern North Island.
61. The nikau reaches its southern limit on the Chatham Islands at 44°S.
62. Greenwood, R. M. and Atkinson, I. A. E. 1977. 'Evolution of the divaricating plants of New Zealand in relation to moa browsing.' *Proc. N. Z. Ecol. Soc. 24:* 21-33.
63. Cockayne, L. 1928. 'The Vegetation of New Zealand — Die Vegetation der Erde XIV.' Engelmann, Leipzig (Ed. 2).
64. McKelvey, P. J. 1963. 'The synecology of the West Taupo indigenous forests.' *N. Z. Forest Service Bull. 14.* Government Printer, Wellington.
65. McKelvey, P. J. 1973. 'The pattern of the Urewera Forests.' Forest Research Institute, Technical paper 59.
66. Cameron, R. J. 1955. 'Mosaic or cyclical regeneration in North Island podocarp forests.' *N. Z. Journ. For. 7:* 55-65.
67. Beveridge, A. E. 1973. 'Regeneration of podocarps in a central North Island forest.' *N. Z. Journ. For. 18:* 23-35.
68. Poole, A. L. 1937. 'A brief ecological survey of the Pukekura State Forest, South Westland.' *N. Z. Journ. For. 4:* 78-85.
69. Holloway, J. T. 1954. 'Forests and climates in the South Island of New Zealand.' *Trans. Roy. Soc. N. Z. 82:* 329-410.
70. Wardle P. 1963. 'The regeneration gap of New Zealand gymnosperms.' *N. Z. Journ. Bot. 1:* 301-15.
71. Robbins, R. G. 1962. 'The podocarp-broadleaf forests of New Zealand.' *Trans. Roy. Soc. N. Z. Botany I:* 33-75.
72. Havel, J. J. 1971. 'The Araucaria forests of New Guinea and their regenerative capacity.' *Journ. Ecol. 59:* 203-14.
73. Whitmore, T. C. 1966. 'The social status of *Agathis* in a rain forest in Melanesia.' *Journ. Ecol. 54:* 285-301.
74. Veblen, T. T. and Stewart, G. H. 1982. 'On the conifer regeneration gap in New Zealand: The dynamics of *Libocedrus bidwillii* stands on the South Island.' *Journ. Ecol. 70:* 413-36.
75. Whitmore, T. C. 1975. *Tropical Rain Forests of the Far East.* Clarendon, Oxford.
76. Jones, E. W. 1956. 'Ecological studies on the rain forest of Southern Nigeria IV. The plateau forest of the Okumu Forest Reserve.' *Journ. Ecol. 44:* 83-117.
77. The terrestrial form is probably a distinct species. (See *Eagle's Trees and Shrubs of New Zealand.* Second Series).
78. Bieleski, R. L. 1959. 'Factors affecting growth and distribution of the kauri.' *Aust. Journ. Bot. 7:* 252-94.
79. Esler, A. E. and Rumball, P. J. 1975. 'Gumland Vegetation at Kaikohe, Northland, New Zealand.' *N. Z. Journ. Bot. 13:* 425-36.
80. Kumarahou flowers rubbed together with water form a sort of lather.
81. Cheeseman, T. F. 1896. 'On the flora of the North Cape District.' *Trans. N. Z. Inst. 29:* 333-85.
82. Silver pine is replaced by yellow silver pine (*Lagarostrobus intermedium*) in western Southland and Stewart Island.
83. Mark, A. F. and Smith, P. M. F. 1975. 'A lowland vegetation sequence in South Westland: Pakihi bog to mixed beech-podocarp forest. Part. I: The principal strata.' *Proc. N. Z. Ecol. Soc. 22:* 76-89.
84. Rigg, H. H. 1962. 'The pakihi bogs of Westport, New Zealand.' *Trans. Roy. Soc. N. Z. Botany I:* 91-108.
85. Baylis, G. T. S. 1948. 'Vegetation of Great Island, Three Kings group. *Rec. Auck. Inst. Mus.* 3: 239-52.
86. Oliver, W. R. B. 1948. 'The Flora of the Three Kings Islands.' *Rec. Auck. Inst. Mus. 3:* 211-88
87. Wright, A. E. 1983. 'Conservation status of the Three Kings Islands flora in 1982.' *Rec. Auck. Inst. Mus. 20:* 175-84.

**Chapter 5**

88. Wardle, John A. 1984. *The New Zealand Beeches. Ecology, Utilisation and Management.* N. Z. Forest Service.
89. *Nothofagus* means 'false beech'.
90. Poole, A. L. 1958. 'Studies of the New Zealand *Nothofagus* species III. The entire-leaved species.' *Trans. Roy. Soc. N. Z. 85:* 551-64.
91. Poole, A. L. 1948. 'The flowering of beech.' *N. Z. Journ. For. 5:* 422-42.
92. Preest, D. S. 1963. 'A note on the dispersal characteristics of the seed of the New Zealand podocarps and beeches, and their biogeographical significance.' Pp. 415-24 in *Pacific Basin Biogeography, 10th Pacific Science Congress.* Gressitt, J. T. (Ed.). Bishop Museum Press, Honolulu.
93. Cranwell, L. M. 1939. 'Southern-beech pollens.' *Rec. Auckl. Inst. Mus. 2:* 175-96.
94. Black beech and its relative mountain beech are both tolerant of dry conditions, but can also grow on poorly drained swampy sites. This is not as anomalous as it seems, as poorly drained soils are considered to be 'physiologically dry' as the lack of free oxygen greatly reduces the water absorption efficiency of the roots.
95. Wardle, P. 1965. 'A comparison of alpine timberlines in New Zealand and North America.' *N. Z. Journ. Bot. 3:* 113-35.
96. Wardle, P. 1971. 'An explanation for alpine timberlines.' *N. Z. Journ. Bot. 9:* 371-402.
97. Wardle, P. 1964. 'Facets of the distribution of forest vegetation in New Zealand.' *N. Z. Journ. Bot. 2:* 352-66.

98. Wardle, P. 1980. 'Ecology and distribution of silver beech ( *Nothofagus menziesii*) in the Paringa district, South Westland, New Zealand.' *N. Z. Journ. Ecol. 3:* 23-34.

99. Dansereau, P. 1964. 'Six problems in New Zealand vegetation.' *Bull. Torrey Bot. Club 91:* 114-40.

115. Atkinson, I. A. E. , and Greenwood, R. M. 1980. 'Divaricating plants and moa browsing: a reply.' *N. Z. Journ. Ecol. 3:* 165-66.

116. Lowry, J. B. 1980. 'Evolution of divaricating plants in New Zealand in relation to moa browsing.' *N. Z. Journ. Ecol. 3:* 165.


**Chapter 6**


100. Cockayne, L. 1912. 'Observations concerning evolution, derived from ecological studies in New Zealand.' *Trans. Proc. N. Z. Inst. 44:* 1-50.

101. Cockayne, L. 1899. 'An enquiry into the seedling forms of New Zealand phanerogams and their development. Part II.' *Trans. Proc. N. Z. Inst. 31:* 361-98.

102. A partly comparable situation exists in south-west Madagascar,[103] where in arid, but foggy, coastal sites there is an abundance of twiggy, densely interlaced, small-leaved shrubs belonging to genera from a number of different families. Some of the shrubs are spiny and some have zig-zag branching and short shoots as in New Zealand. The small-leaved shrubs of Madagascar are not related to those of New Zealand but probably derive from tree and shrub relatives with much larger leaves in the tropical rain forests.

103. Koechlin J. , Guillaumet J-L and Morat, Ph. 1974. *Flore et Vegetation de Madagascar.* Vaduz: Cramer.

104. Greenwood and Atkinson distinguish 'scrub' and 'shrubland' thus: '. . . scrub is distinguished from forest by having most stems less than 10 cm d. b. h. Shrubland is distinguished from scrub by having a woody cover of less than 80 per cent.'

105. Allan, H. H. 1924. 'On the hybridity of *Coprosma cunninghamii.* ' *N. Z. Journ. Sci. Tech.* 6: 310-18.

106. Went, F. W. 1971. 'Parallel evolution.' *Taxon 20:* 197-226.

107. Diels, L. 1897. 'Vegetations Biologie von Neu-Seeland.' *Bot. Jahrb. 22:* 202-300.

108. Wardle, P. 1963. 'Evolution and distribution of the New Zealand flora as affected by Quarternary climates.' *N. Z. Journ. Bot. 1:* 3-17.

109. Godley, E. J. 1979. 'Leonard Cockayne and evolution.' *N. Z. Journ. Bot. 17:* 197-215.

110. McGlone, M. S. and Webb, C. J. 1981. 'Selective forces influencing the evolution of divaricating plants.' *N. Z. Journ. Ecol. 4:* 20-28.

111. *Muehlenbeckia astonii* and a few divaricate species of *Olearia* are deciduous.

112. Cain, S. A. , Castro, G. M. de O. et al. 1956. 'Applications of some phytosociological techniques to Brazilian rain forest. Part III. Life-form and leaf size classes.' *Amer. Journ. Bot. 43:* 928-41.

113. Moas were flightless birds, some species of which were larger than emus or ostriches. They became extinct a few centuries before the arrival of Europeans.

114. Burrows, C. J. 1980. 'Some empirical information on the diet of moas.' *N. Z. Journ. Ecol. 3:* 125-30.


**Chapter 7**


117. Molloy B. P. J. , Burrows C. J. et al. 1963. 'Distribution of sub fossil forest remains in eastern South Island.' *N. Z. Journ. Bot. 1:* 68-77.

118. McGlone, M. S. and Moar, N. T. 1977. 'The *Ascarina* decline and post-glacial climatic change in New Zealand.' *N. Z. Journ. Bot. 15:* 485-9.

119. Molloy, B. P. J. 1969. 'Recent history of the vegetation.' Pp. 340-60 in *The Natural History of Canterbury.* A. H. & A. W. Reed.

120. Connor, H. E. and Macrae, A. H. 1969. 'Montane and subalpine tussock grasslands in Canterbury.' Pp. 168-204 in *The Natural History of Canterbury.* A. H. & A. W. Reed.

121. McGlone, M. S. and Bathgate, J. L. 1983. 'Vegetation and climate history of the Longwood Range, South Island, New Zealand, 12000 BP. to the present.' *N. Z. Journ. Bot. 21* : 293-315.

122. McGlone, M. S. 1978. 'Forest destruction by early Polynesians, Lake Poukawa, Hawkes Bay, New Zealand.' *Journ. Roy. Soc. N. Z.* 8: 275-81.

123. Calder, D. M. 1961. 'Plant ecology of subalpine shingle river beds in Canterbury, New Zealand.' *Journ. Ecol. 49:* 581-94.

124. van Steenis, C. G. G. J. 1981. *Rheophytes of the World.* Sijthoff and Noordhoff, The Netherlands.

125. Moore, L. M. and Adams, N. M. 1963. *Plants of the New Zealand Coast.* Paul's Book Arcade, Hamilton.

126. Molloy, B. P. J. and Simpson, M. J. A. 1980. 'Taxonomy, distribution and ecology of *Pachystegia* (Compositae). A progress report.' *N. Z. Journ. Ecol. 3:* 1-3.

127. Betts, M. Winifred. 1918. 'Notes on the autecology of certain plants of the peridotite belt, Nelson. Part 1. Structure of some of the plants. (No. 1).' *Trans. N. Z. Inst. 50:* 230-43.

128. Druce, A. P. , Bartlett, J. K. and Gardner, R. O. 1979. 'Indigenous vascular plants of the serpentine area of Surville Cliffs and adjacent cliff tops, north-west of North Cape, New Zealand.' *Tane 25:* 187-206.

129. Healy, A. J. and Edgar, Elizabeth. 1980. *Flora of New Zealand, Volume 3. Adventive Cyperaceous, Petalous and Spathaceous Monocotyledons.* Government Printer, Wellington.

130. Healy, A. J. 1969. 'The adventive flora of Canterbury.' Pp. 261-333 in *The Natural History of Canterbury.* A. H. and A. W. Reed.

131. Armstrong, J. F. 1872. 'On the naturalised plants of the Province of Canterbury.' *Trans. N. Z. Inst. 4:* 284-90.

132. Cockayne, L. 1967. *New Zealand Plants and their story.*

4th Ed. (Ed. E. J. Godley.) Government Printer, Wellington.

133. There may also be a fairly wide shrubland belt where the beech treeline is locally depressed as a result either (a) of persistent fogginess with consequent light and temperature reduction, or (b) of temperature inversion effects at U-shaped glacial valley heads.

**Chapter 8**

134. Wardle, P. 1977. 'Plant communities of Westland National Park and neighbouring lowland and coastal areas.' *N. Z. Journ. Bot. 15:* 323-98.
135. Burrows, C. J. et al. 1979. 'New Zealand heathlands.' Pp. 339-64 in *Ecosystems of the world. 9A. Heathlands and related Shrublands. Descriptive studies.* Elsevier.
136. Mark A. F. 1980. 'Progress in tussock grasslands research since Cockayne's day.' *Proc. Roy. Soc. N. Z. 108:* 122-42.
137. Mark, A. F. 1974. 'Snow tussock grasslands.' *N. Z. Nature Heritage 35:* 976-81. Hamlyns.
138. Dawson, J. W. and LeComte, J. R. 1978. 'Research on *Aciphylla*. A progress report.' *Tuatara 23:* 49-67.
139. Fisher, F. J. F. 1965. 'The Alpine Ranunculi of New Zealand.' Bull. N. Z. D. S. I. R. 165: 1-192.
140. Given, D. R. 1969. 'A synopsis of infrageneric categories in *Celmisia.* ' *N. Z. Journ. Bot. 7:* 400-18.
141. What is generally called a flower in the Compositae is really a very condensed inflorescence (capitulum) comprising many small flowers.
142. Gibson, Neil and Kirkpatrick, J. B. 1985. 'A comparison of cushion plant communities of New Zealand and Tasmania.' *N. Z. Journ. Bot.* 23: 549-66.
143. Cushion bogs are also found at sea level on the south coast of the South Island near Invercargill.
144. Wilson, H. D. 1978. *Wild Plants of Mount Cook National Park.* Field Guide Publication.
145. Mark, A. F. and Bliss, L. C. 1970. 'The high-alpine vegetation of Central Otago, New Zealand.' *N. Z. Journ. Bot. 8:* 381-451.

**Chapter 9**

146. Wardle, P. 1968. 'Evidence for an indigenous pre-Quaternary element in the mountain flora of New Zealand.' *N. Z. Journ. Bot. 6:* 120-5.
147. Fleming, C. A. 1963. 'Age of the alpine biota.' *Proc. N, Z. Ecol. Soc. 10:* 15-18.
148. Raven, Peter H. 1973. 'Evolution of subalpine and al pine plant groups in New Zealand.' *N. Z. Journ. Bot. 11:* 177-200.
149. Wardle, P. 1978. 'Origin of the New Zealand mountain flora, with special reference to trans-Tasman relationships.' *N. Z. Journ. Bot. 16:* 535-50.
150. Dawson, J. W. 1971. 'Relationships of the New Zealand Umbelliferae.' Pp. 43-61, in *The Biology and Chemistry of the Umbelliferae,* V. H. Heywood (Ed.). Academic Press.
151. Webb, C. J. and Druce, A. P. 1984. 'A natural intergeneric hybrid, *Aciphylla squarrosa* x *Gingidia montana,* and the frequency of hybrids among other New Zealand apioid Umbelliferae.' *N. Z. Journ. Bot. 22:* 403-11.
152. Webb, C. J. 1986. 'Breeding systems and relationships in *Gingidia* and related Australasian Apiaceae.' Pp. 383-399, in *Flora and Fauna of Alpine Australasia: Ages and Origins.* Bryan A. Barlow. (Ed.). C. S. I. R. O. , Canberra, Australia.
153. Given, David R. 1973. '*Damnamenia* gen. nov. A new subantarctic genus allied to *Celmisia* Cass. (Astereae — Compositae).' *N. Z. Journ. Bot. 11:* 785-96.

**Chapter 10**

154. Taylor, B. W. 1955. 'The Flora, Vegetation and Soils of Macquarie Island.' *Aust. Nat. Res. Exp. Rep. Ser. B. (II), Botany:* 1-192
155. Seppelt, R. D., Copson, G. R. and Brown, M. J. 1984. 'Vascular flora and vegetation of Macquarie Island.' *Tasmanian Naturalist* 78: 7-12.
156. Gillham, M. E. 1967. *Sub-antarctic Sanctuary. Summertime on Macquarie Island.* Gollancz, London.
157. *Puccinellia macquariensis* was also thought to be endemic, but has recently been discovered on Campbell Island.
158. Wace, N. M. 1960. 'The botany of the southern oceanic islands.' *Proc. Roy. Soc. Lond., Ser. B, 152:* 475-90.
159. Some seeds can float in the sea and arrive unharmed on distant shores. One such was picked up on the coast of Macquarie and when germinated in Australia turned out to be *Caesalpinia bonduc,* widespread in tropical coastal habitats. It could not have survived on Macquarie, but at least it got there.
160. Meurk, Colin D. 1982. 'Supplementary notes on plant distributions of the subantarctic Auckland Islands.' *N. Z. Journ. Bot. 20:* 373-80.
161. Johnson, P. N. and Campbell, D. J. 1975. 'Vascular plants of the Auckland Islands.' *N. Z. Journ. Bot. 13:* 665-720.
162. Cockayne, L. 1909. 'The ecological Botany of the subantarctic islands of New Zealand.' Pp. 182-235 in *The Subantarctic Islands of New Zealand,* Vol. 1. Government Printer, Wellington.
163. Godley, E. J. 1965. 'Notes on the vegetation of the Auckland Islands.' *Proc. N. Z. Ecol. Soc. 12:* 57-63.
164. Oliver, R. L. and Sorensen, J. H. 1951. 'Botanical investigations on Campbell Island.' *Cape Exped. Ser. Bull. No. 7.* D. S. I. R., Wellington.
165. Meurk, C. D. 1982. 'Regeneration of subantarctic plants on Campbell Island following exclusion of sheep.' *N. Z. Journ. Ecol. 5:* 51-58.
166. Williams, G. R. 1982. 'Species-area and similar relationships of insects and vascular plants on the southern outlying islands of New Zealand.' *N. Z.*

*Journn. Ecol. 5:* 86-96.

167. Cheeseman, T. F. 1909. 'On the systematic Botany of the islands to the south of New Zealand.' Pp. 389-471, in *The Subantarctic Islands of New Zealand,* Vol. 2. Government Printer, Wellington.
168. Wright, A. C. S. 1959. 'Soils of Chatham Island.' *N. Z. Soil Bureau Bull.* 19.
169. Cockayne, L. 1902. 'A short account of the plant covering of Chatham Island.' *Trans. N. Z. Inst. 34:* 243-325.
170. Kelly, G. C. 1983. 'Distribution and ranking of remaining areas of indigenous vegetation in the Chatham Islands, with site notes and introductory text.' Dept. of Lands and Survey Land Resources Inventory map sheet with extended legend.
171. Given, David R. and Williams Peter A. 1984. *Conservation of Chatham Island Flora and Vegetation.* Botany Division, D. S. I. R.
172. There is, however, a large endemic ground *Astelia, A. chathamica,* which grows in moist sites in the south of the main island.
173. Sykes, W. R. 1977. Kermadec Islands Flora. 'An annotated check list.' *D. S. I. R. Bull. 219:* 1-296.
174. Oliver, W. R. B. 1910. 'The Vegetation of the Kermadec Islands.' *Trans. N. Z. Inst. 42:* 118-75.
175. Laing, R. M. 1915. 'A revised list of the Norfolk Island flora with some notes on the species.' *Trans. Proc. N. Z. Inst. 47:* 1-39.
176. Turner, J. S. Smithers, C. W. and Hoogland, R. D. 1968. *The Conservation of Norfolk Island.* Australian Conservation Foundation. Special Publication Number I.
177. Oliver, W. R. B. 1896. 'The vegetation and flora of Lord Howe Island.' *Trans. Proc. N. Z. Inst. 49:* 94-161.
178. Rodd, A. N. and Pickard, John 1983. 'Census of vascular flora of Lord Howe Island.' *Cunninghamia 1:* 267-80.
179. Hutton, Ian, 1986. *Lord Howe Island.* Conservation Press, Canberra.
180. Green, P. S. 1970. 'Notes relating to the floras of Norfolk and Lord Howe Islands, I.' *Journ. Ann. Arb. 51:* 204-20.
181. Green, P. S. 1979. 'Observations on the phyto-geography of the New Hebrides, Lord Howe Island and Norfolk Island.' Pp. 41-53 in *Plants and Islands,* D. Bramwell (Ed.). Academic Press.

**Chapter 11**

182. Beadle, Noel C. W. 1981. *The Vegetation of Australia.* Cambridge Univ. Press.
183. Groves, R. H. (Ed.). 1981. *Australian Vegetation.* Cambridge Univ. Press.
184. Webb, L. J. 1959. 'A physiognomic classification of Australian rain forests.' *Journ. Ecol.* 47: 551-70.
185. Harris, Thistle Y. 1970. *Alpine Plants of Australia.* Angus and Robertson.
186. Johns, R. J. 1982. 'Plant Zonation.' Pp. 309-30 in *Biogeography and Ecology in New Guinea.* Vol. 1, L. Gressitt (Ed.). Junk.
187. Wardle, P. 1973. 'New Guinea: our tropical counterpart.' *Tuatara* 20: 113-24.
188. Schmid, Maurice, 1981. *Fleurs et Plantes de Nouvelle-Calédonie.* Les Editions du Pacifique.
189. Morat, Ph., Veillon, J. -M., and MacKee, H. S. 1986. 'Floristic relationships of New Caledonian rainforest phanerogams.' *Telopea* 2: 631-79.
190. Morat Ph., Jaffre, T., Veillon, J.M., and MacKee, H. S. 1986. 'Affinités floristiques et considérations sur l'origine des maquis miniers de la Nouvelle-Calédonie.' *Bull. Mus. natn. Hist. nat., Paris, 4e ser., 8, sect. B, Adansonia:* 133-82.
191. Godley, E. J. 1960. 'The botany of southern Chile in relation to New Zealand and the Subantarctic.' *Proc. Roy. Soc. Lond. B., 152:* 457-75.
192. McQueen, D. R. 1976. 'The ecology of *Nothofagus* and associated vegetation in South America.' *Tuatara* 22: 38-68.
193. Moore, D. M. 1975. 'The alpine flora of Tierra del Fuego.' *Anal. Inst. Bot. Cav. 32:* 419-40.
194. Goodspeed, T. Harper, 1950. *Plant Hunters in the Andes.* Robert Hale, London.
195. Wace, N. M. 1965. 'Vascular Plants.' Pp. 201-66 in *Biogeography and Ecology in Antarctica,* van Dye, P. and van Mieghem, J. (Eds.). Junk.

**Chapter 12**

196. Drake, Hilary and Burrows, C. J. 1980. 'The influx of potential macrofossils into Lady Lake, north Westland, New Zealand.' *N. Z. Journ. Bot. 18:* 257-74.
197. Pocknall, D. T. 1980. 'Modern pollen rain and Aranuian vegetation from Lady Lake, north Westland, New Zealand.' *N. Z. Journ. Bot. 18:* 275-84.
198. Kemp, Elizabeth M. 1978. 'Tertiary climatic evolution and vegetation history in the southeast Indian Ocean region.' *Palaeogeography, Palaeoclimatology, Palaeoecology 4:* 169-208.
199. Crook, Keith A. W. 1981. 'The break-up of the Australian-Antarctic segment of Gondwanaland.' Pp. 3-14 in *Ecological Biogeography of Australia,* A. Keast (Ed.). Junk.
200. Dettmann, Mary, E. 1981. 'The Cretaceous flora.' Pp. 357-75 in *Ecological Biogeography of Australia,* Allen Keast (Ed.). Junk.
201. The family Araucariaceae is still represented in New Zealand but with only one species, *Agathis australis* (kauri). Macrofossils of *Araucaria* have also been discovered, but as pollen of this genus and *Agathis* is very similar it is unclear when *Araucaria* became extinct in New Zealand.
202. Martin, Helene, A. 1981. 'The Tertiary flora.' Pp. 393-426 in *Ecological Biogeography of Australia,* Allen Keast (Ed.). Junk.
203. Smith, J. M. B. (Ed.) 1982. *A History of Australasian Vegetation.* McGraw-Hill.

204. Fleming, C. A. 1979. *The Geological History of New Zealand and its Life*. Auckland University Press.
205. Mildenhall, D. C. and Pocknall, D. T. 1984. 'Palaeobotanical evidence for changes in Miocene and Pliocene climates in New Zealand.' Pp. 159-71 in *Late Cainozoic Climates of the Southern Hemisphere,* J. C. Vogel (Ed.). Sasqua International Symposium. A. A. Balkema, Netherlands.
206. Pocknall, D. T. and Mildenhall, D. C. 1984. 'Late Oligocene — early Miocene spores and pollen from Southland, New Zealand.' *N. Z. Geol. Surv. Paleontol. Bull. 51.*
207. Campbell, J. D. 1985. 'Casuarinaceae, Fagaceae, and other plant megafossils from Kaikorai Leaf Beds (Miocene), Kaikorai Valley, Dunedin, New Zealand.' *N. Z. Journ. Bot. 23:* 311-20.
208. McGlone, M. S. 1985. 'Plant biogeography and the late Cenozoic history of New Zealand.' *N. Z. Journ. Bot. 23:* 723-49.
209. Webb, L. J., Tracey, J. G. and Jessup, L. W. 1986. 'Recent evidence for autochthony of Australian tropical and subtropical rainforest floristic elements.' *Telopea* 2: 575-89.

# Index